(Syn)aesthetics: *Re*defining

Also by Josephine Machon

PERFORMANCE AND TECHNOLOGY:
Practices of Virtual Embodiment and Interactivity
(*co-edited with Susan Broadhurst*)

SENSUALITIES/TEXTUALITIES & TECHNOLOGIES:
Writings of the Body in 21st Century Performance
(*co-edited with Susan Broadhurst*)

(Syn)aesthetics: *R*edefining Visceral Performance

Josephine Machon

© Josephine Machon 2009, 2011

All rights reserved. No reproduction, copy or transmission of this publication may be made without written permission.

No portion of this publication may be reproduced, copied or transmitted save with written permission or in accordance with the provisions of the Copyright, Designs and Patents Act 1988, or under the terms of any licence permitting limited copying issued by the Copyright Licensing Agency, Saffron House, 6–10 Kirby Street, London EC1N 8TS.

Any person who does any unauthorized act in relation to this publication may be liable to criminal prosecution and civil claims for damages.

The author has asserted her right to be identified as the author of this work in accordance with the Copyright, Designs and Patents Act 1988.

First published in hardback 2009
First published in paperback 2011 by
PALGRAVE MACMILLAN

Palgrave Macmillan in the UK is an imprint of Macmillan Publishers Limited, registered in England, company number 785998, of Houndmills, Basingstoke, Hampshire RG21 6XS.

Palgrave Macmillan in the US is a division of St Martin's Press LLC, 175 Fifth Avenue, New York, NY 10010.

Palgrave Macmillan is the global academic imprint of the above companies and has companies and representatives throughout the world.

Palgrave® and Macmillan® are registered trademarks in the United States, the United Kingdom, Europe and other countries.

ISBN 978–0–230–22127–7 hardback
ISBN 978–0–230–33690–2 paperback

This book is printed on paper suitable for recycling and made from fully managed and sustained forest sources. Logging, pulping and manufacturing processes are expected to conform to the environmental regulations of the country of origin.

A catalogue record for this book is available from the British Library.

A catalog record for this book is available from the Library of Congress.

10 9 8 7 6 5 4 3 2 1
20 19 18 17 16 15 14 13 12 11

Printed and bound in Great Britain by
CPI Antony Rowe, Chippenham and Eastbourne

For Andrew

Contents

List of Figures	x
Acknowledgements	xii
Notes on Interview Contributors	xiii
Epigraph	xviii
Preface to the Paperback Edition	xix
Introduction: Redefining Visceral Performance	1

Part 1

1.1	Defining (Syn)aesthetics	13
	Synaesthesia – defining the parameters of (syn)aesthetics	15
	Engaging *sense* with sense	16
	From the neurological to the theatrical	19
	Imagination, the ineffable and embodied knowledge	21
	Live performance	24
	Tracing a feminized style	26
	Interdisciplinary, intercultural practice and the (syn)aesthetic hybrid	29
	'New Writing' and the visceral-verbal	31
	Demanding a new discourse	32
1.2	Connecting Theories	34
	Nietzsche's Dionysian	35
	The Russian Formalists – Dionysian disruptions in linguistic play	37
	Barthes – *jouissance* and pleasurable texts	39
	Kristeva's *semiotic chora* and genotext	40
	Cixous and Irigaray – *écriture féminine*	42
	Artaud – disturbance and sensation in the Theatre of Cruelty	44
	Novarina – corporeality and carnage in the Theatre of the Ears	46
	Barker – imagination and disturbance in the Theatre of Catastrophe	48

	Broadhurst and the liminal – touching the edge of the possible	50
	(Syn)aesthetics – connecting theories	52
1.3	(Syn)aesthetics in Practice	54
	The (syn)aesthetic hybrid – a 'total' (syn)aesthetic	55
	The performing body and the (syn)aesthetic style	62
	Disturbing speech patterns – the visceral-verbal *play*text	69
	(Syn)aesthetics – *re*defining visceral performance	80

Part 2

	Introduction: A (Syn)aesthetic Exchange	85
2.1	Felix Barrett and Maxine Doyle of Punchdrunk: In the Prae-sens of Body and Space – The (Syn)aesthetics of Site-Sympathetic Work	89
2.2	Lizzie Clachan and David Rosenberg of Shunt Theatre Collective: A Door into Another World – the Audience and Hybridity	100
2.3	Akram Khan: The Mathematics of Sensation – the Body as Site/Sight/Cite and Source	112
2.4	Marisa Carnesky: Trapping the Audience in the Fantasy – Instinct, the Body and the Magic of the Experiential	124
2.5	Naomi Wallace and Kwame Kwei-Armah: Desire, the Body and Transgressive Acts of Playwriting – on Writing and Directing *Things of Dry Hours*	132
2.6	Linda Bassett: Bypassing the Logical – Performing Churchill's *Far Away*	144
2.7	Jo McInnes: A Text That Demands to Be Played With – Performing Kane's *4.48 Psychosis*	153
2.8	Graeae's Jenny Sealey and Playwright Glyn Cannon: Seeing Words and (Dis)Comfort Zones – the Fusion of Bodies, Text and Technology in *On Blindness*	160
2.9	Sara Giddens and Simon Jones of Bodies in Flight: The In-Betweens, Where Flesh Utters and Words Move – on Flesh, Text, Space and Technologies	172

2.10 Leslie Hill and Helen Paris of Curious:
     Embodied Intimacies – On (the) Scent,
     Memory and the Visceral-Virtual   184

Notes   197
Bibliography   205
Index   217

Figures

1.1 Punchdrunk's *Sleep No More* (2003). © Punchdrunk: L-R Geir Hytten, Robert McNeill, Elizabeth Barker, Hector Harkness, Gerard Bell, Dagmara Billon, Mark Fadoula, Andreas Constantinou, Sarah Dowling. Photo: Stephen Dobbie — 13

1.2 Polly Frame and Graeme Rose in Bodies in Flight's *Skinworks* (2002). Photo credit: Edward Dimsdale — 34

1.3 Image from Shunt's *Dance Bear Dance* (2002–3), courtesy of Lizzie Clachan. Reproduced with kind permission of Shunt Theatre Collective. Photo: © Shunt — 54

1.4 Punchdrunk's *Sleep No More* (2003). Performer: Geir Hytten. Photo: Stephen Dobbie. Image © Punchdrunk — 57

1.5 Marisa Carnesky in *Carnesky's Ghost Train* (2008). Photo Credit: Marcus Ahmad — 60

1.6 Leslie Hill in Curious' *On the Scent*. Image © Curious/Arts Admin. Reproduced with kind permission. Photo credit: Hugo Glendinning — 63

1.7 Bodies in Flight. *Deliver Us* (1999–2000). Performers, Mark Adams, Polly Frame. Image © Bodies in Flight — 66

1.8 *Zero Degrees* (2005) Khan with Sidi Larbi Cherkaoui. Reproduced with kind permission of Akram Khan Dance Company. Photo credits: Tristram Kenton — 76

1.9 *Zero Degrees* (2005) Khan with Gormley sculpture. Reproduced with kind permission of Akram Khan Dance Company. Photo credits: Tristram Kenton — 77

2.1 Punchdrunk's *Faust* (2006). L-R: Fernanda Prata and Geir Hytten. Image © Punchdrunk. Photo credit: Stephen Dobbie — 90

2.2 Image from Shunt's *Amato Saltone* (2006), courtesy of Lizzie Clachan. Reproduced with kind permission of Shunt Theatre Collective. Photo © Shunt — 110

2.3 *Sacred Monsters* (2006). Akram Khan and Sylvie Guillem.
     Reproduced with kind permission of
     Akram Khan Dance Company. Photo credit: Mikki Kunttu   122
2.4 Marisa Carnesky's *Jewess Tattooess* (1999–2001). Photo
     credit: Manuel Vasson   125
2.5 Graeae's *On Blindness* written by Glyn Cannon. L-R:
     Steven Hoggett, Jo McInnes, Scott Graham, David Sands.
     Image © Graeae. Reproduced with kind permission   168
2.6 Polly Frame in Bodies in Flight *Skinworks* (2002).
     Photo credit: Edward Dimsdale   173
2.7 Curious *Lost & Found* (2005). L-R: Helen Paris,
     Leslie Hill, Lois Weaver. Image © Curious/Arts Admin.
     Photo credit: Hugo Glendinning   185

Cover Image: Punchdrunk's *Masque of the Red Death* (2007–8).
© Punchdrunk 2008. Photo credit: Stephen Dobbie.
Reproduced with kind permission of Punchdrunk

Back Cover: *bahok* (2008). Akram Khan Company with
the National Ballet of China. Photo credit: Hugo Glendinning.
Reproduced with kind permission of Akram Khan Dance Company

Acknowledgements

I would like to give special thanks to Janet Free for her considerable time, support and expert advice. Special thanks also to Paul Woodward for his ongoing encouragement and professional insight. My grateful thanks to Eleanor Machon for sharing her professional expertise. I am indebted to Felix Barrett, Linda Bassett, Glyn Cannon, Marisa Carnesky, Lizzie Clachan, Maxine Doyle, Sara Giddens, Leslie Hill, Simon Jones, Akram Khan, Kwame Kwei-Armah, Jo McInnes, Helen Paris, David Rosenberg, Jenny Sealey and Naomi Wallace for the interviews they have given. Their generosity of time and spirit has proven vital to this book. For their assistance relating to certain interviews I wish to acknowledge Sven Oliver Van Damme of Akram Khan Dance Company, Colin Marsh of Punchdrunk and Helen Filmer. For permission to use images I am grateful to Akram Khan Dance Company, Bodies in Flight, Marisa Carnesky, Curious and Cheryl Pierce at Arts Admin, Graeae, Punchdrunk and Shunt. I also wish to thank Paula Kennedy at Palgrave for her advice and support throughout the publishing process. Finally, I am thankful for my family and friends for their constant support, love and laughter and to Andrew and Rufus, for everything.

Notes on Interview Contributors

Linda Bassett is an internationally renowned actress. Her extensive theatre history includes roles in Caryl Churchill's *Fen*, *Serious Money* and playing Harper in the original production of *Far Away* (Royal Court/Albery, 2000/1). Recent film credits include *Cass* (Cass Films Ltd), *The Reader* (Reader Films Ltd), *Kinky Boots* (Touchstone/Harbour), *Calender Girls* (Harbour Pictures), and Ella Khan in *East Is East* for which she won Best Actress Award, Senana Internacional de Cine Valladolid Espania, and was nominated Best Actress in the London Evening Standard British Film Awards and BAFTA Awards. Bassett's wide-ranging television credits include playing Queenie in *Larkrise to Candleford* for the BBC.

Bodies in Flight was formed in 1990 by choreographer Sara Giddens and writer Simon Jones. They make interdisciplinary performance work that has at its heart the contemporary, everyday experience of being caught between the physical and psychological self; the encounter between flesh and text where words move and flesh utters. Giddens is a choreographer, teacher, lecturer, researcher and project manager. Her choreography seeks to challenge both the boundaries of conventional dance and performance and how those works can be documented and disseminated. Jones is a writer and scholar. He has taught theatre studies at Lancaster University, and is currently a reader in performance at the University of Bristol. He has been a visiting scholar at Amsterdam University (2001), a visiting artist at The School of the Art Institute of Chicago (2002), and worked with Spell#7 Performance (Singapore) in a revival of his performance text *Beautiful Losers* (1994/2003).

Glyn Cannon is a British playwright. His plays include; *The Kiss* (Hampstead Theatre) and *Gone*, a modern adaptation of Sophocles' *Antigone*, first produced at the Pleasance Courtyard for the 2004 Edinburgh fringe festival and winner of a Fringe First award from The Scotsman. Also, *Nebuchadnezzar* first produced in 2002 at the Latchmere theatre, Battersea, London. His play *On Blindness* was produced by Paines Plough, Graeae Theatre Company and Frantic Assembly (Soho Theatre, London, 2004). He was Associate Playwright of Paines Plough, 2003–4, and is an associate artist of The Miniaturists.

Marisa Carnesky is an Olivier award-winning performance artist and show-woman who has presented performance events internationally, such as her solo shows *Jewess Tattooess* and *Magic War*. She was also an original collaborator and performer on *C'est Duckie*. She is creator, director and producer of *Carnesky's Ghost Train* which is inspired by her Eastern European heritage and her work as the AHRC Fellow of the National Fairground Archive at the University of Sheffield.

Curious was formed in 1996 by Leslie Hill and Helen Paris. The company has a reputation for edgy, humorous interrogations of contemporary culture and politics. Curious have produced many projects in a range of disciplines including performance, installation, publication and film such as *Vena Amoris*, *On the Scent*, *Lost & Found be)longing* and *Autobiology*. Curious' investigations involve intimate, personal journeys alongside public research and enquiry which leads to collaborations and conversations with a huge range of people such as truckstop waitresses, biological scientists, political refugees, ocularists, nuclear weapons experts, sex workers, old folks' social groups and lost property workers. Hill and Paris, as Curious, are Artsadmin artists.

Akram Khan is an acclaimed choreographer and artistic director of Akram Khan Dance Company, set up in 2000. He began dancing at the age of seven and studied with the Kathak dancer and teacher Sri Pratap Pawar. At the age of 14, he was cast in Peter Brook's legendary production of *Mahabharata*, appearing in the televised version of the play broadcast in 1988. Company works include *Kaash* (2002) a collaboration with artist Anish Kapoor and composer Nitin Sawhney, *ma* (2004), for which he received a South Bank Show Award (2005); *zero degrees* (2005), a collaboration with dancer Sidi Larbi Cherkaoui, sculptor Antony Gormley and composer Nitin Sawhney, nominated for a Lawrence Olivier Award in 2006 and *Sacred Monsters* (2006), a collaboration with Sylvie Guillem, with additional choreography by Taiwanese choreographer Lin Hwai Min. Khan was invited by Kylie Minogue in summer 2006 to choreograph a section of her 'Showgirl' concert. In collaboration with the National Ballet of China and dancers from his own company, Kahn created *bahok* (2008). Khan's collaboration with Juliette Binoche, *in-i*, premiered at The National Theatre, London, in September 2008.

Kwame Kwei-Armah is a British actor, broadcaster, columnist, director and playwright. Kwei-Armah was writer-in-residence at the Bristol Old Vic from 1999–2001. He is currently writer-in-residence for BBC Radio Drama, an associate artist at the National Theatre of Great Britain and at CENTERSTAGE, Baltimore. His triptych of plays set in the habits of the

African-Caribbean community, *Elmina's Kitchen*, *Fix Up*, and *Statement of Regret*, premiered at the National Theatre between 2003 and 2007, with *Elmina's Kitchen* transferring to London's West End, Baltimore and Chicago. He has won the Evening Standard Charles Wintor Award for Most Promising Playwright and the RECON Community Leadership Award, 2007 and has been nominated for a Lawrence Oliver Award and a BAFTA. He recently received an Honorary Doctorate from The Open University. At Baltimore's CENTERSTAGE, he directed MacArthur Fellowship recipient Naomi Wallace's *Things of Dry Hours* in 2007 and has recently directed his own play, *Let There Be Love* at the Tricycle Theatre, London.

Jo McInnes has performed in theatre, film and television and currently runs her own production company. She was in the original Royal Court production and revival of Sarah Kane's *4.48 Psychosis*.

Punchdrunk is the UK's leading exponent of immersive theatre. Based in London, the company produces original works of installation performance on an epic scale. Punchdrunk's most recent productions, *Faust* (2006) and *The Masque of the Red Death* (2007) set box office records for site-located theatre, taking art form and audiences to the next frontier of experiential practice. Felix Barrett is artistic director of Punchdrunk and a graduate and Honorary Fellow in Drama, University of Exeter. Barrett has conceived, designed and directed all of Punchdrunk's productions since founding the company in 2000. He is a Critics' Circle Drama Award winner (Best Design, *Faust*, 2006) and one of the first recipients of a Paul Hamlyn Foundation Breakthrough Fund award (2008). Maxine Doyle trained at Roehampton and the Laban Centre and was Artistic Director of dance theatre company First Person (1996–2002). Touring work for the company included *Plastic Chill* (1999–2000) and *It's Only a Game Show* (2002). Doyle began her association with Punchdrunk in 2003. She has co-directed *Sleep No More* (2003), *The Firebird Ball* (2004), *The Yellow Wallpaper* (2005), *Faust* (2006) and *Masque of the Red Death* (2007–8). Doyle was a semi-finalist with Barrett for The Place Prize (2006) and created the site responsive piece *Picnic for Dancing City* (2007).

Jenny Sealey is artistic director of Graeae, a disabled-led theatre company that profiles the skills of actors, writers and directors with physical and sensory impairments. The artistic approach creates aesthetically accessible productions that include a disabled and non-disabled audience. It is the leading British company in its field, and has been led by Sealey since 1997. Her productions with Graeae include: *Peeling* by Kaite O'Reilly, *Diary of an Action Man* by Mike Kenny (with Unicorn, Time Out's Critics Choice 2003); *On Blindness* by Glyn Cannon (with Paines Plough and Frantic Assembly) and a voice/BSL audio described and ethnically diverse production of *Bent* by Martin Sherman (2004). Her production of *Blasted* by Sarah Kane toured nationally in spring 2006 and sold out in London's Soho Theatre in January 2007. *Static* by Dan Rebellato, in collaboration with Suspect Culture, toured the UK from February to May 2008.

Shunt is a collective of ten artists creating large-scale performance events in unexpected, abandoned or derelict buildings throughout London. Shunt's founder members are Serena Bobowski, Gemma Brockis, Lizzie Clachan, Louise Mari, Hannah Ringham, Layla Rosa, David Rosenberg, Andrew Rutland, Mischa Twitchin and Heather Uprichard. Their current home is a sprawling labyrinth of railway arches under London bridge station, the site of shunt's last two productions (*Tropicana* and *Amato Saltone*) and of The Shunt Lounge since September 2006. Lizzie Clachan's Shunt designs include *Contains Violence* (Lyric Hammersmith), *Amato Saltone* (with National Theatre), *Tropicana* (with National Theatre), *Dance Bear Dance* and *The Ballad of Bobby François*. Clachan's other design credits include *Shoot/Get Treasure/Repeat* (Paines Plough), *On Insomnia and Midnight* (Royal Court), *Julie* (National Theatre of Scotland), and *Lady Bird* (Royal Court). David Rosenberg graduated from the University of Manchester, Faculty of Medicine (MB ChB) and was a Senior House Officer in Anaesthetics. For Shunt his directing responsibilities include *Dance Bear Dance*, *The Ballad of Bobby Francois* and *The Tennis Show*. He has recently received a NESTA Fellowship to explore the role of the audience in live performance.

Naomi Wallace was born in Kentucky, and presently lives in North Yorkshire, England. Wallace's major plays include *Things of Dry Hours*, *One Flea Spare*, *The Trestle of Pope Lick Creek*, *In the Heart of America*, *Slaughter City*, *The War Boys*, *The Inland Sea*, *Birdy* (an adaptation for the stage of William Wharton's novel) and *The Fever Chart: Three Visions of the Middle East*. Wallace's work has been produced internationally and has been awarded the Susan Smith Blackburn Prize, the Fellowship of Southern Writers Drama Award, the Kesselring Prize, the Mobil Prize, an NEA grant, a Kentucky Arts Council Grant, a Kentucky Foundation for Women grant and an Obie Award for best play. Wallace is also a recipient of the MacArthur Fellowship. Her award-winning film, *Lawn Dogs*, was produced by Duncan Kenworthy. Her film *The War Boys* (adapted from her play of the same name, with Bruce Mcleod) will be released in 2009. Forthcoming productions in 2009 include her new play, *The Hard Weather Boating Party*, at Actor's Theatre (Humana Festival) and *Things of Dry Hours* at New York Theatre Workshop. Wallace is presently under commission by The Public Theater, Tricycle Theatre, Clean Break of London and the Oregon Shakespeare Festival.

intoxicated by it overwhelmed by it delighted by it intrigued by it he speaks intimately as if only to me yet also always to those I'm dimly aware are around me all of us *present* aware that they're responding in the same way laughing listening breathing with me the gentleness of his voice enticing my ears into the narratives he weaves the electric moments as his harnessed body becomes Jean Cocteau spinning and falling breathtakingly through space as he lands in an embodied shift of character as Miles Davis shooting up heroin in a shadowplay projection the sensation of the needle on skin penetrating me through the subtlest rhythm of the shadowy movement feeling the fusion of imagesoundspace the architectural wizardry the play between characters pleasurable and tangibly textured takes me into my body head and stomach not separate immediate innate inhabited the delicate balance between whirring technology and subtle physicality taking me into this narrative drawing me in and beyond the hushed phone calls and cacophonous soundtracks and more and more all fusing together to delight and disturb my senses telling me stories that convey philosophical thought so I feel them I touch them so that I can hear raindrops form without yet dropping as I leave the theatre and so that I want to write and write and talk and talk on it in it to think around the ideas and the aesthetic yet also revel reve(a)l solely in my embodied experience aware that my reaction is at once tactile (as he falls through the air head and guts suffer vertigo, diving in the locomotion of his body in the elongated soundscore) and esoteric (mind expands overturns alights on a higher plane the space of it opens out as it enters the beautifully worked through canon of connections that exist narratively philosophically emotionally in each shift of the story) from the undergraduate student talking talking it through with classmates and tutor to here now today these same responses flood over me as I recall as I recall...

(Machon on Robert Lepage's *Needles and Opium,* 1993)

Preface to the Paperback Edition

I write this preface at a time when experiential theatre and 'immersive' events are at the forefront of both populist and highbrow culture. From installation art to live performance to underground cinema events to music festivals to blitz parties and burlesque nights out, the focus on direct participation with a rekindling of the human senses is paramount; creating *an experience*, emphasising the visceral, is the paradigm. With this in mind, it feels timely that this new paperback edition of *(Syn)aesthetics: Redefining Visceral Performance* should arrive.

'(Syn)aesthetics' is a term that defines a theory in/as/of practice. If new to this book or returning to it, I recommend that the reader navigate the more theoretical material in the first part of the book via the interviews in Part 2. With their focus on live performance, alongside a commitment to sensory and immersive processes in creating work, the discussions in Part 2 articulate how (syn)aesthetics works in practice across diverse forms.

Whether reading the book cover-to-cover or cross referencing the theory of Part 1 with the conversations in Part 2, I would emphasise from the outset that the fusion of the reflective and the immediate within the tone of each interview is intended to draw the reader into a deeper understanding of how (syn)aesthetics exist within and beneath the sensual analysis of these dialogues. The reader can make these associations across Parts 1 and 2, uncovering the specific terminology where it exists in my contributions to the conversations or by drawing connections between the quintessential features of (syn)aesthetics and the artists' own ways of analysing and describing their work. It is my hope that the reader experiences (syn)aesthetics as a sympathetic discourse that is both inspired by and located firmly within the work itself. With this in mind, an important aim of this book is to present a theory that might encourage students in this field to enjoy and give authority to their own, individual analysis of such work.

Part 1 of this book sets out the notable characteristics of the (syn)aesthetic performance style and its accompanying mode of analysis. This section of the book surveys a variety of scientific, critical and performance theories to underpin and elucidate the parameters of its own analytical approach. It draws on a linguistically rich frame of reference via *all* of the theorists to which it refers, including those from scientific disciplines such as Richard E. Cytowic and A. R. Luria. The appropriation of the terms and features of the neurological condition of synaesthesia is an engaging central element of this artistic theory. The language of the analysis owes much to the potent and evocative definitions and descriptions provided by diagnostic research into this phenomenon. Furthermore, the delightfully disruptive nature of the critical theories employed within Part 1, from Friedrich Nietzsche to Luce Irigaray, serves to support and elucidate essential features of (syn)-aesthetic practice and analysis and highlights the linguistically playful and formalistically shapeshifting nature of the theory itself. In addition to this, the application of Howard Barker and Valère Novarina, as performance practitioners theorising in this sphere, highlights the performative nature at the heart of (syn)aesthetic analysis.

Following this, and in retrospect, as I reread this book I note the many tics and traits that I employ, which exploit parenthesis, emphasis and homonymity to extend this playful approach to 'meaning-making'. The intent throughout is to foreground the ludic fusions integral to this analytical approach. Playfulness aside, the overall style of Part 1 favours a more traditional academic frame in order to make this introduction to (syn)aesthetics, as a creative discourse for analysis/practice, accessible to a wide audience, which includes undergraduates in this field. Part 2 presents conversations with leaders in the field of performance, which displays the key features of the (syn)aesthetic style, to illustrate the preceding theory. These artists articulate directly a sensate approach that is at the heart of their practice, thereby validating any sensual response that arises from participation in the performance event. The combined purpose of Parts 1 and 2 is to articulate a new and immersive critical methodology in which the artist, the audience, and the academic become equals in their approach and response to such work.

In whatever way the reader negotiates their own trajectory through Parts 1 and 2, attention should be drawn (and is reiterated throughout) to the mutually beneficial exchange between performance and analysis. This analytical attitude is intended to empower the student reader on the quest to understand, articulate and relish the way in which critical thinking exists within and around every aspect of the artistic process in their chosen field of study. Early in the Introduction I draw attention to the organic development of this theory of (syn)aesthetics to highlight how it has come about as a direct result of my own interests as a student of Drama through to my current practice and teaching. As I write, I am excited by the increasing innovations in large scale immersive practice and intimate and ritualised one-on-one performance; I remain inspired by the continuing poetic and provocative experiments in 'New Writing' for theatre and am still compelled by the eloquence of the body in contemporary performance practice. This wealth of tantalising theatrical experiences available to us today is exactly the type of work that originated my desire to articulate my own embodied response to such work; a response where often 'words' could not do the experience justice. My attitude to analysis continues to evolve, shaped by the visceral experience at the core of such practice and I believe that the approach that I have taken in this book embraces this within its very form, resolutely heeding Susan Sontag's advice:

> Interpretation takes the sensory experience of the work of art for granted, and proceeds from there. This cannot be taken for granted now.... What is important now is to recover our senses. We must learn to *see* more, to *hear* more, to *feel* more (1982: 104, emphasis original).

Josephine Machon
April, 2011

Introduction: *Re*defining Visceral Performance

> Creating performances and writing about those performances require acts of critical and creative imagination; both contend with the imperatives carried by 'the act'.
>
> (Peggy Phelan, 1998: 7)

Since the late twentieth century a performance style has emerged which exploits diverse artistic languages to establish an 'experiential' audience event via the recreation of visceral experience.[1] Impossible to define as a genre due to the fluidity of forms explored, this performance style places emphasis on the human body as a primary force of signification and utilizes the ever-increasing possibilities in design and technology. In addition to this it has reclaimed the verbal as a visceral act and embraced the written word, a factor previously denied in certain performance practice where the language of physicality has been prioritized. Consequently, this style enables practitioners and audience members alike to tap into pre-linguistic communication processes and engages with an awareness of 'the primordial' via such sensually stimulated perception. Merged with this is the potential to engender a certain feeling of transcendence, of comprehending ideas, experiences and concepts in a unique way. As a result, this style produces a response of disturbance that can be simultaneously challenging and exhilarating, at once unsettling and pleasurable.

To attempt to define it as a specific genre in contemporary performance proves difficult. Arguably it can be traced through ancient, ritualized practice such as Noh Theatre, Kathakali, Greek tragedy; through Shakespearean and Jacobean theatre, to the avant-garde practice of Jacques Copeau, Vsevolod Meyerhold, Antonin Artaud, Isadora Duncan, Samuel Beckett, Jacques Lecoq or Martha Graham (to name but few); and onwards to the innovators of the late twentieth century through to the present such as Pina Bausch or Robert Lepage. This suggests that the style itself is not new and attempts to articulate it are not new. Arguably, such a style draws widely on these ancient modes of theatre and inherits the forms and techniques of certain late twentieth century performance practice. In particular, transgressive female practice in the arts, specifically from the late 1960s onwards; intercultural, interdisciplinary ensemble work which became internationally prominent from the 1980s and through to the present and the developments of a visceral play-writing aesthetic established, in Britain in particular, throughout the 1990s. In short, by defining this as a particular contemporary style, an integral feature of such practice is that whilst pushing forward the boundaries of performance to explore contemporary experience it returns to existing conventions that draw on the unique power of ritual in performance.

Such performance practice demands a change in the criteria of appreciation, as it is at loggerheads with analytical methods previously applied to the intellectual and literary style of theatre production prominent in much theatre of the twentieth century. Performance work that is sensate and transgressive in its very form can produce a response in the individual audience member that goes beyond the discourse of critical analysis as it stands.[2] The problem of articulating experiences that are, on the whole, 'inarticulable', arises due to the fact that the act of immediate perception is primarily located in the body. Secondly, this style of performance may bring about the forementioned transcendental quality, which is also difficult to put into words. The immediacy of such a response, at once corporeal and arcane, has to greatly influence subsequent processes of interpretation.

My own experience of being caught in and between analysis and practice began as an undergraduate student following my individual experience of Robert Lepage's *Needles and Opium* in 1993. From this defining moment and throughout my years as a student of performance I have explored theories which go some way to marrying thinking with practice. In addition to the theorists referenced below and in the following chapters, I happily alighted on the writings of Peggy Phelan, who has done much to prioritize the embodied nature of the human experience of arts practice in general, alongside dance theorists, such as Sondra Horton Fraleigh, who went some way to addressing the issue of how analysis meets practice in performance theory. Yet all of this thinking placed work by genre or discipline and tended to separate ideas around the moving body and the written text.

My quest to find a sympathetic mode of analysis for performance that is idiosyncratically visceral and fuses disciplines, rather than fitting into one form or genre, became more urgent from 1997 onwards, working with undergraduate Drama students. Here there was a mutual attempt to find vocabulary and to explore theories that 'did justice' to the quality of experience we had had as audience members in relation to a variety of works. These included Theatre de Complicite's *Street of Crocodiles*, Caryl Churchill's *The Skriker* and De la Guarda's *Villa! Villa!*, alongside archive video material of Samuel Beckett's *Not I*, Steven Berkoff's *Metamorphosis*, Pina Bausch's *Bluebeard* and DV8's *Dead Dreams of Monochrome Men*. We became aware of a sense of being in-between the work itself and our reflections on that work; an ability to recall the various visceral responses we had in an embodied way (including to the video footage) yet, as a result of the very nature of this experience, a certain inability to articulate our response within an academic setting. During this time, the organic development of my thinking around this area also emerged from my own interdisciplinary practice collaborating with dancers and physical performers in the exploration of the relationship between writing (my own and others') and dance; between the body, text and space.

As this illustrates, it is an ongoing, individual confrontation with such a style in general and throughout my own practice, that has led to this investigation into the methodological gaps that exist in current performance analysis. This book seeks to address the dilemma of verbally analyzing experiential performance work in order to foreground the significance of this style in theatre – historically, culturally, and in terms of academic study and individual interpretation. Finding the 'correct' words to define a response can be challenging and frustrating but is ultimately pleasurable, providing the strategies to distinguish, discuss and further explore the impulse of such a style and the mode of appreciation that it affects. The genesis of ideas and much of the key terminology used throughout this book have arisen directly from this urge to find a discourse for experiential performance events, which articulates both the approach to practice as well as the methods of appreciation that occur in the experience of that work, for practitioners and audience members alike.

Parts 1 and 2 of this book identify the need for an interpretative device that conjures such a style and endeavours to liberate, through analysis, the varied processes of the appreciation strategy rather than reducing the work as it is discussed.[3] Such a theoretical discourse must itself be, primarily, of the body. It must also provide specific, yet open, vocabulary that acknowledges and encompasses the corporeality of the response.

In order to provide a mode of performance analysis that defines the *full* appreciation process that occurs (that is, from an immediate, individual response to any subsequent intellectual interpretation that transpires), I have exploited the idea of exchange and adaptability and played with the space and the slippage in-between theory and practice in an attempt to fuse both within a mutually sympathetic discourse. It is in the endeavour to find vocabulary which defines and embraces fused corporeal and cerebral experiences that I have adopted the term '(syn)aesthetics' (from 'synaesthesia', the Greek *syn* meaning 'together' and *aisthesis* meaning 'sensation' and 'perception'). I have developed the term in my teaching and research since 2000 to define an interpretative device which describes *simultaneously* a performance style – its impulse, and processes of production – *and* the appreciation strategy necessary to articulate a response to such work.[4] As I will go on to explain in Part 1, it is important to stress that by adopting and adapting the term, I hope to foreground how both the performance style and the appreciation it affects have an integral feature of disturbance; a disturbance that can be unsettling and/or exhilarating. My intention is to assert that the '(syn)aesthetic style' identifies a particular, exciting contemporary performance practice that has grown in recent years, and '(syn)aesthetics' an equivalent interpretative device to analyse such work.

As clarified in Part 1, (syn)aesthetics embraces performance work which constantly resists and explodes established forms and concepts. Consequently, (syn)aesthetics is always open to developments in contemporary practice and analysis. (Syn)aesthetic work shifts between performance disciplines, just as it shifts between the sensual and intellectual, the somatic, ('affecting the body' or 'absorbed through the body') and the semantic (the 'mental reading' of signs). As a result (syn)aesthetic practice can be understood to have a certain 'shapeshift' morphology, its only constant being the fused somatic/semantic manner of its performance style and subsequent audience response.[5]

Within the diverse texts of (syn)aesthetic practice, there are three key performance strategies, arising in no order of priority, which are peculiar to the (syn)aesthetic performance style. In brief they are; the (syn) aesthetic hybrid, which is a special manipulation of the *gesamtkunstwerk* (a term coined by Richard Wagner meaning 'total art work'); a predominance of the actual body as text in performance; and an unusual rendering of writerly speech to establish a visceral-verbal *play*text.[6]

'Play' as foregrounded in my emphasis in *play*texts is fundamental to the impulse in creation and appreciation of the (syn)aesthetic style, and particularly to my intentions for the written and verbal texts surveyed in this book. Such play is inherently ludic. By that I intend that it is subversive, curious and imaginative. Developing Immanuel Kant's *'free play* of imagination' (1911: 58–60, emphasis original) summarized by Susan Broadhurst as, a 'pleasure that depends ... on consciousness of the harmony of the two cognitive powers imagination and understanding' (1999a: 28), this concept of 'play' is discussed in more detail in Part 1, Chapters 1–3 and is a key feature of the discussions in Part 2.[7]

At this point I must provide some disclaimers. Firstly, in identifying these three central strategies, discussed at length in Chapter 3, I do not intend to imply that (syn)aesthetic performance *always* incorporates all three. As the consideration of the (syn)aesthetic hybrid emphasizes, the performance style consists of diverse combinations of texts. Following this, within any (syn)aesthetically styled performance there can be a fusion and slippage between the dominance of any one of the three strategies under scrutiny here.

Secondly, as regards the interpretative device of (syn)aesthetics, my aim is to engender an open discourse that embraces visceral artistic practice and allows for the symbiotic exchange that *must* occur between performance and any theory that seeks to articulate and define it. In responding to performance work which resists closure, so too does the (syn)aesthetic mode of appreciation and analysis resist closure. I should stress that this method of analysis is applicable to the experience of any (syn)aesthetically styled work from the fine arts and music as well as a variety of technological media, including certain cinema and television works, such as the film and television work of David Lynch, Peter Greenaway and Derek Jarman, or the films of Andrew Kotting or Alejandro Jodorowsky, as discussed by Helen Paris and Leslie Hill, and Marisa Carnesky, in Part 2. In this way this style and process of appreciation is applicable to the arts *in general*. In order to remain true to the focus of this project and to emphasize the breadth of interdisciplinary practice already present in the realm of 'theatre' the focus of this book will remain rooted in live performance practice.

In explaining (by way of liberating) the visceral, experiential quality of the (syn)aesthetic style, I emphasize from the outset the central features within interpretation of 'corporeal memory' and 'embodied knowledge'. These provide an intuitive knowledge that refers human perception back to its own primordial, or *chthonic* (from the Greek, 'of, or to, the earth') impulse. What I intend by 'corporeal memory' is that the sensate external body both produces and appreciates its own 'language' in performance. The 'language' of the performing body alongside the visceral impact of any other sensual element of the performance work is experienced by the audience through the traces of this language in our own flesh; both the external tactile flesh and the internal viscera.[8] This 'internal' encompasses the emotional *and* the physiological or sensational capabilities of the physical body. Work invested with such a quality has the potential to appeal to an equivalent chthonic sensibility within audience reception that allows for the slippage between the human faculties of intellectual and instinctual perception.[9] This factor alone can affect a certain disturbance in the processes of production and appreciation in (syn)aesthetics.

A final request is that the reader accepts that (syn)aesthetics is a heterogeneous mode of analysis which both supports and 'explains' a continually morphing and mutable performance style. In this respect it is a playful and open theoretical device. As Part 1: Chapter 1 highlights, like the performance style it scrutinizes, (syn)aesthetic theory, and the terms employed therein, serve to resist definition in the very act of defining. As a result, it is my intention to make a performance discourse available to that work which has previously been difficult to name in critical theory in general and performance theory specifically due to the inexplicable quality integral to the innate, *individual* response during performance and appreciation. The purpose is to provide a theoretical term that encompasses a non-linguistic, instinctive, intertextual and intersemiotic mode of interpretation and analysis.[10] As emphasized in Chapter 1, the term (syn)aesthetic is incumbent of all these features in its etymology.

Quintessential features of the physiological condition of synaesthesia served to inspire this theory of (syn)aesthetics. Drawing on the experiences of this medically defined neurological condition, extensively from the research of Richard E. Cytowic and A.R. Luria in particular, alongside V.S. Ramachandran and E.M. Hubbard, (syn)aesthetics serves to define that which is inexpressible and firmly based in the holistic interpretative capabilities of the human body.

As the following chapter outlines, my thinking, in accordance with the neurocognitive theories I turn to, accepts that synaesthetics is an actual, proven, sensory experience. That said, I reinforce from the outset, the terms of my analysis *adopt and adapt* terminology and frames of reference from the scientific study of synaesthesia in order to establish a discourse for the experiential qualities of the arts in general and the performance work under consideration in particular. My intention is to appropriate certain ideas from neuroscientific research by way of explaining a stylistic approach in performance and a *quality of experience* in appreciation, including any subsequent affective interpretation, that it produces. This book is not a performative study of the condition of synaesthesia, nor is it an overview of how synaesthesia is present or applicable to the arts.[11] Instead, it is an appropriation of significant thinking from neurocognitive research adapted to describe the parameters of (syn)aesthetics as a new performance theory. I should stress, the playful use of parenthesis reinforces the presence of the fused 'aesthetics' of performance as much as the fused nature of visceral perception.

As mentioned, an important feature of this book is the fact that the theory, the terms used to explain the experience of the work, have been generated by the performance work itself. In this way, I have attempted to find what Susan Sontag argues for; that is, theory which reveals the 'sensuous' nature of form (1982a: 103). Many of the ideas and terms assigned focus primarily on an audience perspective of what it is to experience this work in a fused corporeal and cerebral fashion. With this in mind it must be emphasized that a crucial feature of (syn)aesthetics as an appreciation strategy is that it prioritizes an immediate, individual and innate response to work.

Having provided the framework and overview for (syn)aesthetics as the defining discourse that simultaneously describes and interprets the work in Part 1, the intention with Part 2 is to allow a variety of practitioners to speak for themselves by way of illustrating the nature of the work and the analytical approaches it demands. Fascinating insight is offered in these discussions which consider the work of Bodies in Flight, Curious, Graeae, Punchdrunk and Shunt and of the performance artist Marisa Carnesky and choreographer Akram Khan. The writing practice of Caryl Churchill and Sarah Kane is contemplated by Linda Bassett and Jo McInnes respectively and that of Naomi Wallace by the writer herself with Kwame Kwei-Armah. With these conversations, the aim is not to enforce a use of terminology in reflective practice, nor to encourage the practitioners to employ key terms unnecessarily. Instead, Part 2 aims to illustrate the ideas at work in a more organic fashion and to encourage the reader to draw those analytical connections for her or himself. Primarily, the focus of the discussion in Part 2 is to facilitate the artists in their articulation of the experiential quality that is at the heart of their performance methodologies and to reveal the sensuous approach that stimulates and surrounds the work under reflection.

Part 2 both validates and celebrates the sensual response akin to (syn)aesthetic work and supports the argument for such a critical methodology. It also provides a space where the diversity of this performance style is made clear and where professional practitioner, audience member and scholar are equal in their analysis of the work, sharing embodied responses on a number of levels. The tone of the discussion goes some way to illustrating the emphasis on playfulness within the approach to performance as much as to interpretation, and highlights a certain performative quality in discursive practice. In this way Part 2 seeks to bring together these experiences within a continuum of mutually evolving appreciation.

As is apparent in this introduction, throughout this book there is a preponderance of notions of slippage, exchange and play in all areas of this study; a pleasurable revelling in the potential of boundaries blurred; an enjoyment of the possibilities within the 'both and' as opposed to the 'not, but'. In particular the slippage, exchange and play between theories, style, form and content as well as that between production and appreciation. Of the latter, I aim to ensure that the pleasure to be taken in blurring the place where strategies of appreciation start and processes of production end is accentuated. This continuum of theory, practice and individual, immediate appreciation is of great significance to this argument. The focus on the exchange between theory and practice, the verbal and physical, the corporeal and intellectual underpins my use of the term (syn)aesthetics.[12] This immediately sensory approach, exploiting the hybridity of thinking/producing/writing/receiving is actively explored in Part 2 in the discussions with the practitioners. Consequently the book as a whole emphasizes how practice is fundamental to theory and individual experience is fundamental to analysis.

In Part 1, (syn)aesthetics surveys a variety of scientific, critical and performance theories to support and present the terms of its own analysis. As critical tools these are employed and applied solely to explain, describe and illustrate the (syn)aesthetic style in both production and appreciation. Briefly referencing ideas in the thinking of Immanuel Kant, Jacques Derrida, Maurice Merleau-Ponty and Elaine Scarry in Chapter 1, I then go on to condense and evaluate the philosophical and critical theories of Friedrich Nietzsche, The Russian Formalists, Roland Barthes, Julia Kristeva, Hélène Cixous and Luce Irigaray, alongside the performance theories of Antonin Artaud, Valère Novarina, Howard Barker and Susan Broadhurst, in order to show how they support and elucidate (syn)aesthetics. Chapter 2 connects these critical and performance perspectives to emphasize the union of the senses in human perception and certain critical analysis and to highlight the significance of embodied knowledge and subversive play in artistic practice. The theories serve to accentuate quintessential features of the (syn)aesthetic impulse in production and appreciation. In particular they highlight the notion of immediacy in analysis and place great emphasis on the body and embodied knowledge as the primary force of interpretation.[13] In selecting these particular critical approaches an intention is to foreground how (syn)aesthetics does not push the potential of linguistic practice to the background nor does it focus on verbal practice as the foremost language in performance.

Overall the purpose of this book is to provide an insightful introduction to the exchange between analysis and practice in (syn)aesthetics in a comprehensive and accessible manner. (Syn)aesthetics combines the artistic principle of (syn)aesthetics (literally, fused aesthetics), marrying the interdisciplinary with the intersensual in artistic terms, with characteristics of the physiological condition of synaesthesia (the neurological condition involving a fused sensual perception) within the appreciation process. It highlights the sensual exchange that can occur between diverse performance languages; verbal, corporeal, visual, aural, technological and so on. Furthermore, in stressing the importance of theorizing from the work itself, where the impulse to analyse comes directly from the visceral quality of the live performance moment, the experience of the audience in interpreting such open and complex performance texts becomes key. Ultimately, (syn)aesthetics identifies the pleasures and challenges for practitioners and audience alike of presenting, accessing, negotiating and interpreting such work.

Part 1

1.1
Defining (Syn)aesthetics

> Whoever says feeling also says intuition, that is, direct knowledge, inverted communications enlightened from within. There is a mind in the flesh, but a mind as quick as lightning. And yet the agitation of the flesh partakes of the mind's higher matter.
> (Artaud, 1978: 166)

Figure 1.1 Punchdrunk's *Sleep No More* (2003). © Punchdrunk: L-R Geir Hytten, Robert McNeill, Elizabeth Barker, Hector Harkness, Gerard Bell, Dagmara Billon, Mark Fadoula, Andreas Constantinou, Sarah Dowling. Photo: Stephen Dobbie

(Syn)aesthetics derives from 'synaesthesia' (the Greek *syn* meaning 'together' and *aisthesis*, meaning 'sensation' or 'perception'). Synaesthesia is also a medical term to define a neurological condition where a fusing of sensations occurs when one sense is stimulated which automatically and simultaneously causes a stimulation in another of the senses. An individual may perceive scents or words for certain colours, or a word as a particular smell, or experience tastes as tangible shapes. So in terms of this neurocognitive condition, synaesthesia is defined as the production of a sensation in one part of the body resulting from a stimulus applied to, or perceived by, another part. Also, the production, from a sense-impression of one kind, of an associated mental image of a sense-impression of another kind.

In addition to this, I employ within this portmanteau word the definition of 'aesthetics' as the subjective creation, experience and criticism of artistic practice.[1] Following all of these definitions, from the Greek through the scientific to the artistic, my intention is to fuse ideas held within the medical term with those surrounding the aesthetics of performance practice. My reworking of the term as **'(syn)aesthetics'** (with a playful use of parentheses) **encompasses both a fused sensory perceptual experience and a fused and sensate approach to artistic practice and analysis.** My use of parentheses is also intended to distinguish this performance theory from the neurological condition from which it adopts certain features.

As the emphasis of the parentheses suggests, my appropriation of the term aims to foreground various notions of slippage and fusing together. These are; the fusing of separate disciplines within the artistic process; the fusing of this performance practice with a special individual aesthetic appreciation; various fusions of sensory experience within this aesthetic appreciation, combining cerebral and corporeal perception; and the fusing of performance practice with critical analysis. Fused here also reaffirms the 'fused' experience of the human body, an holistic entirety – physiological, intellectual, emotional – thus prioritizing a connection of body and mind within experience.[2] With this play on fusions, (syn)aesthetics provides a discourse that defines simultaneously the impulse and processes of production *and* the subsequent appreciation strategies which incorporate reception and interpretation. (Syn)aesthetics is an aesthetic potential within performance which embraces a fused sensory experience, in both the process and the means of production, as it consists of a blending of disciplines and techniques to create an interdisciplinary, intertextual and 'intersensual' work.

Characteristic of the (syn)aesthetic performance style is its consolidation of a variety of artistic principles, forms and techniques, manipulated in such a way so as to fuse the somatic and the semantic *in order to produce* a visceral response in the audience. The (syn)aesthetic style allows the explicit recreation of sensation through visual, physical, verbal, aural, tactile, haptic and olfactory means.[3] Here, I do not simply refer to the mere description of a sensual experience but *sensation itself* being transmitted to the audience via a corporeal memory, the traces of lived sensate experience within the human body, activated within the perceiving individual. This **fusing of sense (semantic 'meaning making') with** *sense* **(feeling, both sensation and emotion) establishes a double-edged rendering of making-sense/*sense*-making and foregrounds its fused somatic/semantic nature.**[4]

Synaesthesia – defining the parameters of (syn)aesthetics

[T]he senses...become directly in their practice theoreticians.
(Karl Marx qtd. in Taussig, 1993: 98)

To clarify fully the process of audience appreciation, as well as elucidate traits of the (syn)aesthetic style, I draw heavily on quintessential features of the neurocognitive condition of synaesthesia. This is a neurological complication where there is a crossover between the senses. The condition of synaesthesia can be understood, literally, as the fusing of certain senses, often coupled with a combining of cognition and consciousness. This means that a synaesthete experiences a fusion in sensual perception, the most common being where words or letters are perceived as certain colours. Other forms of synaesthesia include the experience of sound as colour or tastes as tangible shapes. Synaesthetes also have extraordinary powers of perception and memory. Synaesthesia defines a human capacity for perception which shifts between realms; between the sensual and intellectual; between the literal and lateral.

Investigations into synaesthesia were first reported by Francis Galton in the nineteenth century (see 1880a, 1880b, 1907). A small number of studies continued into the early twentieth century, such as Luria's thorough documentation of an individual case (1969), most of which treated synaesthesia as a curiosity in both psychological and neurocognitive research. In recent years extensive research has been conducted into synaesthesia that provides clear evidence to support this as an actual physiological condition and sensory phenomenon that is primarily *experiential* and *affective* (see Cytowic, 1994, 1995, 2002; Harrison, 2001; Ramachandran and Hubbard, 2001a, b, 2003, 2005 and van Campen, 2008).

Interestingly, the majority of current research in the area presents strong arguments for synaesthesia being present and active in all human perception from birth but, whereas the majority of humans filter this out and learn to separate sensual experience, only a minority retain this unusual perceptual ability. The process of isolating sensation within perception and analysis is somewhat artificial, the product of learning to distinguish between the senses in order to simplify experience. We generally accept that taste cannot be disassociated from smell for example and when it is it is a reductive experience, yet we rarely acknowledge or celebrate the potential of fused sensual experience in other areas of interpretation. As Cytowic puts it, all humans are synaesthetic but 'only a handful of people are consciously aware of the holistic nature of perception' (Cytowic, 1995: 8) and Ramachandran and Hubbard argue that 'we all have some capacity' for synaesthetic perception (2003: 58).[5] It is this holistic nature of perception that is important to (syn)aesthetics. From this research it is possible to infer that there is the potential for each of us to retain a synaesthetic memory and an ability to relocate this fused perceptual awareness with a given trigger, such as that offered by certain types of artwork. Following this, in (syn)aesthetic appreciation of experiential performance work individual audience members are enabled to reconnect with a (latent) synaesthetic potential. By this I intend that a quality of perception is activated and *felt*, affecting both perception and cognition in the immediate moment, the traces of which are rekindled in the corporeal memory of any subsequent recall and analysis. Here then the potential of the corporeal memory to influence interpretation becomes paramount.

(Syn)aesthetics adopts and adapts terminology and frames of reference from these scientific studies of the condition of synaesthesia in order to clarify and frame a new critical discourse for defining and interpreting the arts in general and the performance work under consideration in particular. It is the theories of Cytowic and Luria, alongside Ramachandran and Hubbard's research into synaesthesia and language development, that are most useful to this theory for contemporary aesthetics. The quintessential diagnostic features these researchers identify define the quality of experience undergone by the audience when appreciating the work in the immediate moment and in subsequent processes of recall and analysis.

Engaging *sense* with sense

Cytowic's diagnostic conclusions are invaluable as they help to define the experiential quality of the audience response (see 1994: 76–7, 2002: 67–70). First, the sensations experienced are involuntary, they cannot be suppressed but are elicited, and the intensity can be influenced by the situation they occur in, usually with some emotional resonance. Second, the sensations can result in a highly emotional response, drawing on the *noetic* (from the Greek *nous* meaning 'intellect' or 'understanding'), a 'knowledge that is experienced directly' which can provide 'a glimpse of the transcendent' (78). The noetic has an 'ineffable quality' in that it makes manifest a complex experience that defies explanation, most simply understood as 'the "a-ha" of recognition' (121, 229). The **'ineffable' is greatly significant to (syn)aesthetic appreciation in that it defines 'that which by definition cannot be put into words'** (119). I expand on this detail further below.

Cytowic details how synaesthetic experiences can be both distracting and difficult to cope with and can also cause ecstasy and be viewed as an achievement. Important to note is that synaesthesia is 'an additive experience' where the combination of senses creates a more complex experience for the perceiver allowing a 'multisensory evaluation' (92, 167, emphasis original). Furthermore, the experiential nature of synaesthesia that evidences 'the force of intuitive knowledge' is crucial in affirming how immediate, personal experience 'yields a more satisfying understanding than analysing what something "means"' and accepts, celebrates even, 'other kinds of knowing' (Cytowic, 1995: 7, 14).

Luria documents how 'synaesthetic sensations' produce states within an individual where 'there is no real borderline between perceptions and emotions' and sensations are 'so vague and shifting it is hard to find words with which to convey them' (1969: 77). Particularly interesting as regards my focus on the potential of written and verbal text in the (syn)aesthetic performance style is Luria's consideration of language as a physical, defamiliarized and sensational act that draws on the powers of the imagination – hearing, appreciating, interpreting and understanding words as rich sensual images – which enables their corporeal (re)perception and recall. Luria records how the interpretation of words 'synaesthetically (determining meaning, that is, through both sound and sense)' ensures that the '*experience* of words' is 'a measure of their expressiveness' (91, emphasis added). This then is a linguistic communication that induces an embodied and imagistic word perception and interpretation. **To experience synaesthetically means to perceive the details corporeally.** Luria details how a synaesthete's extended reference, unlike 'usual' word perception which means that individuals ignore 'the phonetic elements of words' in favour of a primary concern with 'meaning and usage' (86), can ensure the meaning of words is reflected in the aural, visual, tactile and haptic resonance they embody.

Of great significance to (syn)aesthetics is Luria's highlighting of the power of the imagination within a synaesthetic response. Synaesthetic imagination has the ability to 'induce changes in somatic processes' and disrupt 'the boundary between the real and imaginary' (138, 144). Whereas most individuals have in place 'a dividing line between imagination and reality', in those who experience synaesthesia this borderline has 'broken down' (144). This condition thus engages a perceptual faculty that experiences concepts which the majority can 'only dimly imagine' with a palpability that 'verge[s] on being real' (96). Here the synaesthete's experience inhabits 'two worlds at once, like being half awake yet still anchored in a dream' (Cytowic, 1994: 119). Accordingly, within a synaesthetic reaction a somatic, imagistic response can dominate the semantic as 'images begin to guide one's thinking, rather than thought itself being the dominant element' (Luria, 1969: 116). Luria highlights the fact that this play with the imagination allows 'transition to another level of thought' (133), corroborating Cytowic's notion of noetic capabilities in synaesthetic perception.

Cytowic and Luria highlight the impact of 'hypermnesis' to synaesthetic perception. This defines mental reminisces that reactivate the object or experience of the original perception with affective clarity (see Cytowic, 2002: 103–11; Luria, 1969). Put simply, the memory of previously perceived moments is palpable. This is fundamental to (syn)aesthetic interpretation where *the original visceral experience remains affective in any subsequent recall.* Following this any semantic or intellectual analysis that follows is influenced by this affective state, the analysis and articulation of that analysis is invested with that rich and felt quality of experience.

Cytowic and Luria's diagnostic features are not only useful in explaining the quality of experience within the appreciation of (syn)aesthetic work, as made clear below, they also serve to clarify certain features of the performance style, reinforcing my emphasis on the slippage and exchange between practice and analysis. The (syn)aesthetic performance style is concerned with harnessing the full force of the imagination and in breaking down boundaries between the 'real' and the imaginable. It uses graphic images, palpable forms and visceral words to (re)present ideas and experiences. Of absolute relevance is the insistence on verbal language as a corporeal, defamiliarized and sensate act. Significant here is the way in which the neurological condition of synaesthesia provides experiential evidence that words, both on the page and spoken, have the potential to be perceived in a new and exhilarating way. With the visceral-verbal *play*texts of the (syn)aesthetic style the word is defamiliarized and has to be (re)cognized and made sense of via a sensate fusion of verbal and non-verbal means. Following this, within a (syn)aesthetic appreciation process a certain semanticizing of the somatic experience of words occurs during and/or following a performance, where the 'meaning' of the words is reflected in the sound and the *feeling* they embody.

It is here that Ramachandran and Hubbard's research into the visceral quality of linguistic communication in synaesthetic perception is useful, especially in relation to the development of creativity in general and of metaphor in verbal language in particular (see 2001b, 2003).

Ramachandran and Hubbard present a strong case for synaesthetic perception being the norm for all early humans and thereby the impetus for language development (see 2001b, 2003), substantiating the fact that innate, primordial roots exist in modern day, visceral powers of communication and certain perceptual and cognitive processes. Interestingly, Ramachandran and Hubbard also make a case for 'synkinaesia' ('joined movement'), the cross-activation of two motor maps rather than two sensory maps, linking the physicality of the human body and language development (see 2001b: 19–21). What these studies show is that kinaesthetic experience, its cerebral analysis and verbal articulation are conjoined. Cytowic's research corroborates these findings, providing a compelling argument for the embodied synaesthetic nature of language development in general and metaphor in particular from direct physical action and experience (Cytowic, 2002: 276–93). This lends further support to the (syn)aesthetic reclamation of the verbal as a visceral and physical act in certain performance work.[6]

From the neurological to the theatrical

These diagnostic features of the physiological condition encapsulate the quality of experience undergone by the audience during (syn)aesthetic performance. In the most intense cases (syn)aesthetic interpretation comes close to theories surrounding drug induced synaesthesia where a temporary synaesthetic perceptual experience is achieved via the influence of an external/internal hallucinogenic force (see Cytowic, 2002: 100–2, 132 and van Campen, 2008: 103–14). Bassett makes such a reference in her discussion in Part 2 in relation to Churchill's work; 'it's like taking a mind-expanding drug'. Arguably then the experience of (syn)aesthetic work results in, what I would call, artistically induced synaesthesia. To argue that certain performance work can take individual members to a point where they (re)activate their synaesthetic potential is not a fanciful idea inasmuch as Cytowic foregrounds Heinz Werner's arguments that synaesthesia *can* be induced in individuals who are not ordinarily synaesthetic (see Cytowic, 2002: 11).

With more general experiences of (syn)aesthetic work the diagnostic features surveyed above describe the quality of experience and subsequent process of appreciation to be had. Fundamental to such an audience response is 'primitive sensitivity', a 'visual quality of recall' and the experiencing of such work via an 'overall sense' where the somatic response dominates the semantic (Luria, 1969: 28–80). Here then, a 'multisensory evaluation' establishes an 'additive experience' (Cytowic,

1994: 167, 92) within a complex appreciation process. (Syn)aesthetic disturbance defamiliarizes 'known' experience and causes a (re)awakening of a fused cerebral and corporeal memory. It thus has the potential to provide an audience member with a complete corporeal memory in any subsequent processes of recall. (Syn)aesthetic appreciation strategies demand 'a different form of extended reference' (Luria, 1969: 86) in the approach to meaning making which prioritizes giving in to sensation and experience and engaging the critical faculty of the mind, endorsing 'other kinds of knowing' (Cytowic, 1995: 14). This making-sense/*sense*-making factor, fundamental to (syn)aesthetics, is wholly supported by the feature of neurocognitive synaesthesia where 'perceptions without language can have meaning', that is, semantic sense modulates but cannot replace the underlying and prior sensory relations to meaning making (Cytowic following L. E. Marks, 2002: 74). From birth the human capacity for semantic meaning making resides in *experience*, the limbic system being key to the experiential memory required for (re)cognizing and comprehending concepts as much as speech (Cytowic, 2002: 287–90). In short, **synaesthetic cognition describes (syn)aesthetic appreciation in that it is affective and experiential, semantic sense cannot be disassociated from somatic *sense*.**

Also significant to the (syn)aesthetic appreciation process is the breaking down of the boundary between the real and the imaginary to provide a perception of hidden states. The ineffable explains the '(syn)aesthetic-sense', a term coined to define the fusion of cerebral and corporeal cognition within (syn)aesthetic work where the holistic, sentient human body, 'responds with the "a-ha" of recognition' and experiences an 'aesthetic validation that cannot adequately be put into words' (Cytowic, 1994: 229). This highlights a certain dreamlike inhabiting of two states within the appreciation experience of (syn)aesthetic work. **The (syn)aesthetic-sense defines the intuitive human sense that makes sense/*sense* of the unpresentable and the inarticulable.** It is brought about in performance practice where dramatic techniques express ideas, thoughts, emotional experience, psychological states and so on, that are beyond the bounds of conventional communication. As a result the (syn)aesthetic performance style can make the intangible tangible.[7]

Crucial to (syn)aesthetic appreciation is the way in which such a special perception becomes unusual due to the unsettling and/or exhilarating nature of *the process of becoming aware* of the fusion of senses within interpretation. It requires a degree of interpretative (re)cognition by the audience which returns to an innate knowledge, that of emotion over reason, the pre-knowledge of emotional sentience that is peculiar to human consciousness. In this way thinking is disturbed which causes the spectator to see the ideas, experiences and states of the performance *in the moment*, which can force the audience into (re)perceiving the state presented as if for the first time. The effect of such a response can ensure that the individual holds on to the moment they have experienced and remembers this feeling corporeally in any subsequent interpretation of the work, thereby drawing on human powers of hypermnesis. In this way the *experience* of the work is the most important factor in appreciation and impacts on any subsequent intellectual processes of analysis; a visceral cognition via this corporeal memory. It is this fusion of the 'felt' and the 'understood' in making sense/*sense* of intangible, inarticulable ideas that is crucial to (syn)aesthetic appreciation.

For a performance to be wholly (syn)aesthetic there must be this element of disturbance and (re)cognition within appreciation, which can be unsettling, alarming even and/or exhilarating and liberating. Such an experience prioritizes an individual and innate experience in interpretation and lends itself to the notion that when confronted with such work individual audience members experience a sense of **shared ineffability**. This claim is supported in all of the discussions presented in Part 2 of this book.

Imagination, the ineffable and embodied knowledge

Key aspects of (syn)aesthetic comprehension are the imagination, the ineffable and the fact that the work is experienced through the human body. Moving on from scientific research towards supporting aesthetic ideas, Kant, Merleau-Ponty and Scarry help to clarify these aspects.

The ineffable quality of the (syn)aesthetic-sense is further clarified by Kant's theories of 'the sublime', 'a faculty of mind transcending every standard of sense' (Kant, 1911: 91). The sublime draws on the *'free play* of the cognitive faculties' fusing 'imagination' and 'understanding' in a way that has a noetic potential for expressing the inexpressible (see Kant, 1911: 58–60, emphasis original). Kant's sublime further supports the (syn)aesthetic mode of appreciation via the experience of *'negative pleasure'* (Kant, 1911: 91, emphasis original). For Kant the negative pleasure of the sublime comes about through perceptual experience which appears 'to contravene the ends of our power of judgement' and defines that which is 'an outrage on the imagination' (1911: 91). Kant's negative pleasure supports the appreciation experience in (syn)aesthetic work in that it is 'excited...by the imagination in conjunction with the understanding' and 'the sensations' (see Kant, 1911: 120–31). Kant's sublime and negative pleasure helps to clarify the experiential nature of disturbance within appreciation integral to (syn)aesthetic work which 'presents the unpresentable' (Kant, 1978: 35). It points toward the way in which the (syn)aesthetic style, when manipulated to its full, encourages performance to be an experience in its purest definition; to feel, suffer, undergo.

Kant's notion of *'free play'* of imagination and cognition (see 1911: 58–60, emphasis original) is also vital to (syn)aesthetic performance work in general and to writerly practice in particular. Here, as each of the discussions in Part 2 detail (Wallace and Kwei-Armah, McInnes and Bassett in particular), imagination and the imaginative leaps that the audience take with the performance is key to the processes of individual interpretation.[8]

Key to (syn)aesthetics is its concern with a chthonic response to creating and receiving performance, which enables both performer and audience member to tap into primordial, pre-verbal, communication processes. It is this visceral disturbance within (syn)aesthetic appreciation that produces an immediate, affective reading, harnessing the potential of the body to become the experiencing and interpreting agent of the performance. The sensory experience within the (syn)aesthetic response can be more immediate, more tangible than subsequent processes of cerebral analysis, particularly as such cerebral interpretation usually follows the sensory impact. **Fundamental to the (syn)aesthetic response is the notion that the body is the sentient conduit for the appreciation of artistic work in general, and performance in particular.**

With (syn)aesthetic signification and reading, the body produces and interprets a language of the flesh, aided by a corporeal memory. Nietzsche's thinking is fundamental to this idea where the body and its (re)cognitive powers are key to artistic appreciation. For Nietzsche art is 'an organic function' which 'exercises the power of suggestion over the muscles and senses' to reinvigorate mind *as* body, where one 'hears with one's muscles, one even reads with one's muscles' (1968: 426–8). The corporeal memory of the actual body has recollective capabilities that can produce and (re)cognize on an entirely physiological level – a level of appreciation that, by its very nature, challenges linguistic expression. The body generates a wholly sensate form of expression, communicable in its own unique form.

Merleau-Ponty's phenomenological theory is important in theorizing around sensual and embodied perception. Merleau-Ponty argues for the embodied nature of all human perception and being in the world. He destroys the subject/object dichotomy and asserts the human body as a continuum of the natural world and of sensual experience. As regards synaesthetic perception being the norm and fundamental to human experience he corroborates the argument that '[s]ynaesthetic perception is the rule' as 'sensory experience' is achieved with the 'whole body at once' itself 'a world of inter-acting senses' within which 'the experience of the separate "senses" is gained only when one assumes a highly particularized attitude' (see Merleau-Ponty, 2005: 262–75). He highlights the significance of the 'felt' effect of a thing or experience and in doing so supports the primordial basis in which human perception is rooted (244–82).[9]

Scarry validates corporeal capabilities of perception, emphasizing the human body's primordial presence, by arguing that the flesh is 'the sentient source' (1985: 123) which exists both outside and inside of linguistic sign-systems. The performing body has the ability to communicate via the shared human capacity for a corporeal memory, the traces and memories of corporeal experience in the perceiver's body – which incorporates the fused capability of the human body; emotional, physical, sensational, physiological and so on. In this way the human body actuates 'the sharability of sentience' via embodied experience whereby 'having a body means having sentience and the capacity to sense the sentience of others' (326, 233). The sentient human body is 'pre-language' (6), simultaneously asserting and reclaiming a primordial mode of communication. In this way, the holistic, sentient body in the (syn)aesthetic appreciation strategy is crucial in making sense *of* and *from* the senses.

The performing body in the (syn)aesthetic style can be both 'sight' and 'site' of performance.[10] In addition to this, the actual body becomes 'cite' of performance in (syn)aesthetic work, both in the bodies of the performers and those perceiving bodies in the audience, due to the potential it has to affect a corporeal memory in the immediate response, and subsequent processes of recall.[11] This somatic approach to performance, foregrounded in the corporeal, emphasizes the sensuous in terms of performer and audience contact and highlights the 'sharability of sentience' (326) between performer, performance and audience. Furthermore, when the performer's life and bodily experience is used as creative source material, as well as the key performance signifier, as in the work of Carnesky and Kahn, this activates an experiential immediacy in the performance moment where sentient and sensuous sharability enables an embodied knowledge of other(ed) identities and experiences.[12] As a result, embodied knowledge can engage in a unique and actively political way with the marginal and transgressive. This factor is key to the signification of the body in the textual practice of Churchill, Kane and Wallace (see the discussions in Part 2 with McInnes and Wallace and Kwei-Armah in particular in this respect). The 'politics' of sensuous knowledge highlights the potential of the body as the site of performance signification and as the modality for, and cite of, experiential interpretation. This is important in situations where the performances themselves present, as well as produce, a series of sensations which are disturbing in essence because of their visceral impact and demand an appreciation strategy that is firmly based in corporeality.

The making-sense/*sense*-making process that occurs within the (syn)aesthetic performance style asserts an embodied knowledge due to the fusion of corporeal and cerebral perception. This idea of the body as not only a primary signifier but also the principal human instrument that reads in a unique and innate way is of utmost importance to the (syn)aesthetic mode of appreciation. The palpable content of (syn)aesthetic work and the subsequent (syn)aesthetic interpretation is a direct result of such corporeal intervention.

Live performance

Live performance reaches beyond the experience of sensations in the singular due to the fact that it is an amalgamation of all of the senses within a multidimensional, heterogeneous form. Live performance is a medium which can encompass all of the senses in production and reception and thus provide a fused (syn)aesthetic experience. Being a blend of many different artistic impulses, disciplines and techniques (word, movement, design, sound, light, dance, technology and so on), it has the ability to communicate and affect in the greatest sense.

The (syn)aesthetic performance style exploits the experience of the live moment. It explores various combinations of verbal, physical, design and technological texts, and gives predominance to the human body in performance. It presents *play*texts that emphasize *play* in form and are marked by a visceral-verbal quality. Each of these aspects is interrogated in depth in Chapter 3 below. The (syn)aesthetic style in performance has the ability to communicate that which is intangible, in a live(d) and sensate manner, enabling an encounter with ideas as much as with actual presence. It is this that engages a (syn)aesthetic-sense within appreciation.

Of course, an element of (syn)aesthetic performance is the manipulation of form to present various dimensions of sensory qualities found in lived experience. For example, a quality of the aural presented in the visual (a scream as a physical image held), the aural through the oral (a musical melody transposed into a speech pattern), the literary through the physical (the transcription of written data through dance), or the tactile in the aural (the buzz of a needle on the skin translated through music). Furthermore, the (syn)aesthetic style is able to reproduce intangible, sensate experiences via tangible means. For instance, psychological and emotional experience in abstract physical movement which generates the (syn)aesthetic-sense in appreciation. Felix Barrett and Maxine Doyle of Punchdrunk provide eloquent illustration of this in Part 2.

Most significantly, live performance differs from any other artistic medium due to its immediacy, the 'liveness' of the moment. As Phelan asserts, live performance colludes in a continuing, immediate 'interactive exchange' between the work and the audience, where the performers and audience unite in a 'maniacally charged present' (1993: 146–8). As a result the *'presentness'* of sensory experience may be experienced through this immediate witnessing, taking 'present' as 'from prae-sens, that which stands before the senses' (Scarry, 1985: 9, 197, emphasis original). Following this, with (syn)aesthetic performance, the liveness of the performance moment reve(a)ls in the corporeal pleasure of embodied knowledge. In the present moment of a (syn)aesthetic performance, bodily knowledge engages an individual's capacity for a primordial knowing, a pre(sent)-knowing. In this way, a very real exchange of prae-sens and energy between humans exists within the immediacy of live(d) performance. This is a principal feature of the discussion with Barrett and Doyle, Lizzie Clachan and David Rosenberg of Shunt and Hill and Paris of Curious in Part 2.

This is not to argue that any other mode of performance does not exist as 'live' performance or to say that the (syn)aesthetic impulse and process of appreciation is not available with mediated performance. Audio-visual, automated, digital media and biotechnologies in artistic practice may enable experiential perception, particularly those which demand an interactive response and/or are concerned with the exchange between the live(d) experience as a result of technologies employed.[13]

Design and technological aspects in live performance can be manipulated in order to strengthen and foreground the liveness of the live moment. Technology and multimedia design can be interwoven in order to add to the sensate quality of the piece as evidenced in the work of Bodies in Flight, Curious, Graeae and Shunt. The employment of diverse technologies within live performance serves to produce symbiotically compelling performance languages which assert a (re)valuation of live presence in mediatized performance. Hill and Paris provide a compelling argument to this end in Part 2. The (syn)aesthetic response here then, has a great impact, is immediate, intense and powerful because the physical body live in the audience responds to the physical body (alongside additional elements) live in performance, actuating the 'sharability of sentience' (Scarry, 1985: 326). The (syn)aesthetic impulse can be amplified in those performance experiences where there is a direct, visceral connection between the performing and perceiving body in the same space.

The (syn)aesthetic style encompasses a number of genres and elements of the style can be present within a variety of productions. However, it is prevalent in those performances that are on and push forward the boundaries of performance conventions. With this in mind it is useful to survey significant approaches to performance making that saw the emergence of the (syn)aesthetic style.

Tracing a feminized style

Contemporary female performance practice, especially from the late 1960s onwards, created modes of practice, including writing practice, which were resistant to conventional theatre genres and styles, examining and celebrating female artistic modes. Feminized modes are aligned to female physicality with an adherence to the internal and external recurring rhythms of the female physiological, biological and sexual body. Female practitioners also explored alternative conventions of production – site, staging, performance style, writing, devising – in order to resist traditional processes of performance and find a mode that expressed female experience.

There are three areas of female practice that contribute to the (syn)aesthetic inheritance. First, the experimentation with transgressive forms and content which includes active exploration of hybridized practice, incorporating film, video and aural technology into the work, alongside innovative experimentation with writing practice in form and content; second, an explicit use of the body in performance; and lastly, the prioritizing of a discursory position which locates critical theory firmly within artistic practice.

Contemporary female practice in the arts developed new aesthetic forms and strategies, playing with layers of signification and meaning in order to highlight the fragmented form of individual, social and cultural experience. In doing so it became a forerunner for that performance work which explores and expresses 'difference' (and *différance*) in terms of experience and perspectives on reality.[14] Female performance art thus in the very form of the work addressed the theoretical and actual experience of différance. By fusing various disciplines and techniques, such disruptive and disturbing experimental work served to advance writing practice alongside contemporary, feminized (re)workings of the gesamtkunstwerk in order to (re)present female experience and sensibilities.

In addition to the formalistic transgressive acts, female performance practice played with notions of 'disturbance' in terms of content. Female practitioners throughout the 1970s to the present have been particularly keen to explore 'reality' from a feminized perspective, with a willingness to explore taboo personal and political issues via explicit corporeal forms, which influenced wider performance practice.

Following this, the second major thread within female performance practice that is of relevance here is the utilization of the body as stimulus, content, form and site of performance. Female performance instilled an important appreciation of the status of the body *in general* in contemporary performance practice as a mode of exploration and a site for examination. This examination and exploration of the politics and problems of representation of women specific to the women's movement in the arts has had a huge impact on much of today's performance and visual arts practice.

In female practice the explicitly chthonic exploration of corporeality was responsible for finding a physical form for marginal experience and foregrounded a visceral, physical and visual encounter. Female performance was responsible for establishing the body as both site and sight of performance as a result of the use of the self and the body as the content and form of the piece (see Schneider, 1997). The body in performance became a site of potential rather than a fixed given, enforcing a celebration of the multiple 'texts' of the actual body (physical, physiological, cerebral, sexual and so on) rather than merely a merely a passive object in art. Using the body in this way was a direct celebration of the chthonic and primordial integral to human experience (and aligned with the female) and emphasized the abject and sensual within performance. As a result, female explicit body performance demanded a 'sensate involvement' of the audience, exploring the idea of 'eyes which touch' (32). In doing this it established in performance practice complex and highly interrogative physical texts which collapsed the 'distance between sign and signified' and encouraged 'embodied vision' entering a 'feminized domain' of 'seeing beyond the visible' (22–36).

As a direct result of female creative work the body is now not only a stimulus, subject, site and sight for sex-gendered scrutiny and identification but also for the exploration of a complex blend of highly charged mappings – individual, emotional psychological, sexual, historical, cultural, political. Additionally, it becomes the cite of performance in (syn)aesthetic work, engendering a corporeal memory in the processes of immediate and subsequent appreciation. Consequently, in (syn)aesthetic performance the complexity of the actual body is exploited as a versatile conduit for the sending and receiving of performance messages.

The third thread which is of great significance to the (syn)aesthetic style is the fact that female practice upturned strategies for critically appreciating the work presented. Geraldine Harris draws attention to the 'radically ambiguous and "open"' (1999: 49) form of female performance practice and theory which in its ambiguity and playfulness confuses (or defies) categories. Harris emphasizes the feminized questioning of the modes of interrogating performance work which by its very nature 'resist attempts at authoritative, interpretative "mastery"' and as a result evade the appropriation 'to a single, "pure", uncontradictory theoretical position' (21). Harris asserts that the analysis of such work demands a perspectival shift from one theory to another, finding the connections and foregrounding the practice in order to make sense of the practical and theoretical shifts within the performances themselves. In this way, the theoretical tools used when examining the work in question becomes automatically 'self-reflexive in so far as it offers interpretations of performance while questioning the grounds on which these interpretations are constructed' (21). The open and shifting theories proffered in feminized analysis are wholly sympathetic to such creative work.

Rather than theory being separated from practice, feminized analysis, in particular that referred to as *écriture féminine* (Cixous, 1993; Irigaray, 1985) with its focus on embodied writing, located theory firmly within creative practice and moved away from dry and closed styles of theoretical discourse. Female practice and theory has been responsible for changing the way that the field of signs is constructed and read within performance work, stimulating an arena for wider discourses that address différance. Such practice ensured that previously held notions of fixed identities and known gender traits became unfixed and overturned.

The feminized approach, which establishes a supportive theory that continually morphs in order to celebrate the free-play of open and embodied reading, holds the means of responding with both body and mind vital to the appreciation of much performance work. Such a mode of interpretation is fundamental to the (syn)aesthetic performance approach and appreciation strategy which concerns itself with the

Defining (Syn)aesthetics 29

(re)presentation of experience in performance, defamiliarizing signifiers so that the audience interprets previously 'understood' signs anew. The significant factor in certain feminized theories which support arguments for a (syn)aesthetic style is, rather than providing essentialist arguments, they embrace disturbance and transgression, notions of play and pleasure, and the shifting perspectives integral to the experience of différance.

Interdisciplinary, intercultural practice and the (syn)aesthetic hybrid

The prevalence of interdisciplinary, intercultural ensemble companies in Western performance practice in recent years provides an important foundation for the (syn)aesthetic style; in particular the impulse of these companies, such as Tanztheater Wuppertal, Théâtre Bouffes du Nord, Ex Machina and Theatre de Complicite, to interrogate the experiential by fusing performance techniques, pushing forward developments in technological, digital and multimedia possibilities for use in live performance work. This impulse continues in the make up and approach of Akram Khan Company, Bodies in Flight, Curious, Shunt, Punchdrunk and Graeae, Intercultural performance allowed for creative exploration of the diversity of performance 'languages' in order to communicate. Such practice asserted the interlingual (fusing the verbal, physical, technological and so on) potential of performance as a mode of communication.

Interdisciplinary practice was also fuelled by a cultural blurring of boundaries between the arts. The fusion of arts practice from high and low culture has become patently clear in both mainstream and experimental performance. Work that straddles disciplines such as theatre, dance, visual art, virtual realities, on-line gaming, closed-circuit surveillance, opera, pop-music or stand-up comedy is increasingly prevalent and defies categorization. For example, a significant collaborator in the development of the interdisciplinary (syn)aesthetic style, is 'Dance Theatre'. Taking its lead from Bausch's Tanztheater Wuppertal, this interdisciplinary mode combines dance with speech, contrasting everyday gesture and objects with ethereal images and abstract physicality, and blurs aesthetics from opera to cabaret. It has now become a widely explored convention in contemporary dance practice. Interestingly, in doing so, much dance-theatre has returned to the roots of ancient Eastern, African and Asian Dance-Drama where such categorization between disciplines does not exist. Khan draws attention to this fact in Part 2.

In terms of the increase in intertextual and interdisciplinary performance, in recent years it has become increasingly apparent that the theatrical inheritance of early twentieth century practices (for example, Naturalism, Expressionism and Epic Theatre) are being manipulated and reworked to produce a melting pot of styles, techniques, forms, ideologies and conventions. Physical experimentation is indebted to the avant-garde forms of practitioner theorists such as Meyerhold and Artaud as well as to such diverse activities as circus acts, martial arts and bungee jumping. Certain hybridized approaches to interdisciplinary performance that emerged towards the end of the twentieth century, prevalent in the work of practitioners such as Robert Wilson, Laurie Anderson and Pete Brooks, fused the visual, aural, technological and so on, to create a performance aesthetic that met art installation. Such practice went on to influence Carnesky, Shunt and Punchdrunk.

The intercultural/interdisciplinary experimentation from the 1980s onwards provided a wealth of forms and strategies for practitioners to explore, and also demanded new methods of appreciation amongst audiences. It encouraged an instinctive response, allowing the blend of signifiers to work on the audience on a number of levels, ensuring meaning is made from the visceral effect as well as the cerebral impact of the piece. Intercultural/interdisciplinary work demanded a particular sensibility in appreciation and emphasized the fact that such practice does not offer answers or make the journey through the performance easy; instead it opens up questions and embraces the free-play of an individual response.

The intercultural, interdisciplinary experimentation with hybridity is important to the (syn)aesthetic style as it demonstrated an ongoing desire to find new, fused performance languages in order to communicate contemporary human experience. The fact that in recent years intercultural, interdisciplinary practice has been embraced by the mainstream, particularly evident in the success of Ex Machina, Theatre de Complicite, De La Guarda, Shunt and Punchdrunk, suggests that contemporary audiences are excited by work which presents a transgressive blurring of boundaries and that stimulates more than the intellect alone. The free-play of interdisciplinary work is an integral feature of the (syn)aesthetic performance style and the open quality of (syn)aesthetic interpretation. Contemporary performance has developed a (syn)aesthetic sensibility as a result of and in collusion with this recent heritage of intercultural hybridization where the potency of the 'liveness' of live performance is at its strongest.

'New Writing' and the visceral-verbal

Contemporary performance writing has engaged with alternative disciplines and found new ways to communicate influenced by (and influencing) the socio-political and cultural milieu. This has affected an upheaval in the structures, form and content of textual practice in performance. The term 'New Writing', now arguably less popular in theatre parlance than it was in the 1990s, describes plays that are written to express with new verve and passion the concerns of an era, that strive to find new forms and to produce a new stage language which questions and reflects the social, political and cultural mood of an age. In Britain in the mid to late 1990s New Writing was discussed by critics and audiences alike as if it were a new theatre genre. Certain writers including Churchill, Kane and Wallace had developed a style which challenged and disturbed the theatre and its audience.

As Aleks Sierz argues, this writing 'opened up new possibilities' and revived playwriting, 'exploring new areas of expression' by 'suggesting daring new experiments' (2001: xii). It is central to the development of the (syn)aesthetic style in that it established writing *as* sensation that 'jolts both actors and spectators out of conventional responses, touching nerves and provoking alarm' (4). It is an explicit, contradictorily tender and confrontational style that transgresses 'the boundaries of what is acceptable' where 'the use of shock is part of a search for deeper meaning' and a 'rediscovery of theatrical possibility' (5). This movement in playwriting established new structures in writing, revitalizing layout on the page as much as in performance. It embraced brutally poetic language and starkly visceral stage imagery. The propensity to challenge and disturb has defined this style as a new writing aesthetic that cannot be categorized by any one genre and does not employ any one formalistic device. Instead it attempts to embrace and manipulate a variety of dramatic influences in order to explore the fragmented concerns of contemporary experience.

Play texts written with this sensibility demand experimentation in terms of the performance style. Churchill is an emissary for engaging the imagination and breaking down categories by employing dance and music within the layers of her written text. In this way she became a forerunner for a performance-writerly practice that blended forms and disciplines within playwriting. This commitment to formal experimentation created a new performance style that demanded innovation in terms of design, sound, lighting and technology. It can be seen to trace the inheritance of the linguistic and formalistic experimentation from Greek Tragedy through to the Jacobean sensibility and on to Modernist experimentation (writers such as Samuel Beckett and Eugene Ionesco), whilst simultaneously being influenced by immediately current movements in art, music, dance, television, film, digital practice or on-line interaction. It is not just the content of the work but also the form that engages the imagination of the audience, breaking down barriers between traditional theatre writing conventions and new performance potentialities.

These features of new writing are integral to the reclamation of verbal language and the innovations in form and structure in the visceral-verbal (syn)aesthetic *play*text. It is a practice of *play*writing produced from the *play* with the possibilities of live performance. This is apparent in the formalistic experimentation with image, movement and physicality, which is woven into the very fabric of the *play*text itself as evidenced in the writing of Churchill, Kane and Wallace. The transgressive quality of *play*texts in performance encourages practical inventiveness from directors and performers alike. They also demand an immediate and emotionally sentient response from the audience. Such writing is 'experiential, not speculative' with a viscerality that 'forces audiences to react' (4–5) due to the violation of performance expectations in form and content. In the same way (syn)aesthetic *play*writing focuses on the live performance and interrogates the essence of the live(d) theatrical event. It is writing that requires a rich and versatile performance style and asserts a fluid and shapeshifting form that contravenes categorization.

Demanding a new discourse

> Both synaesthesia and the artistic experience are ineffable, and both indescribable by language.
>
> (Cytowic, 2002: 319)

In tracing the inheritance of the (syn)aesthetic style, the different types of experimental work surveyed above can be seen to be driven by a desire to explore and express ideas and experiences in new ways. These approaches fuse conventions, prioritize a shapeshifting free-play and explore and expose diversity in views of reality. Such experimental performance enables practitioners to explore what they know about theatre and to go beyond that, encouraging individuals within an audience to transgress their own boundaries in appreciation of form, content and liveness. This practice demands an analytical approach that merges with the open and perspective-shifting style of the work under scrutiny, as is offered with (syn)aesthetics.

The next chapter plays in the space between theory and practice, where (syn)aesthetics resides. Chapter 2 examines critical and performance theories, finding the connections and fusing them together to elucidate quintessential features of both the performing style and the analytical discourse of (syn)aesthetics.

1.2
Connecting Theories

> We want to hold fast to our senses and to our faith in them – and think their consequences through to the end.
>
> (Nietzsche, 1968: 538)

Figure 1.2 Polly Frame and Graeme Rose in Bodies in Flight's *Skinworks* (2002). Photo credit: Edward Dimsdale

Certain critical and performance theories help to elucidate (syn)aesthetics as a fused mode of practice and appreciation, each with a transgressive and ludic core. All of which prioritize the body as the primary force of interpretation and give credence to the significance of *all* the senses in human perception. Namely, Nietzsche's arguments for a 'Dionysian' artistic impulse; the Russian Formalists theories regarding disruptive linguistics; Barthes' arguments for 'pleasurable text'; Kristeva's theories of transgressive communication; Cixous and Irigarays' *écriture féminine*; and the performance theories of Artaud, Novarina, Barker and Broadhurst. These theories all place emphasis on slippage between the verbal and physical, the chthonic and noetic, the 'felt' and the 'understood'. These connections serve to clarify the impulse underpinning the (syn)aesthetic style, its mode of production, immediacy of appreciation and its requisites in interpretation.

These theories embrace intertextual practice and celebrate the interface between, and flux within, linguistic, corporeal and technological approaches, serving to support (syn)aesthetics as a new form of aesthetic interpretation and the (syn)aesthetic style as an exciting performance mode.

Nietzsche's Dionysian

> The muses of the art of 'illusion' paled before an art that, in its intoxication, spoke the truth.
>
> (Nietzsche, 1967a: 46)

Nietzsche is crucial to understanding the (syn)aesthetic style as his arguments for the Dionysian artistic impulse prove fundamental to the creation of (syn)aesthetic work. The Dionysian is an artistic impulse in opposition to Nietzsche's Apollinian. They are 'interwoven artistic impulses' which, via the tension between them, 'continually incite each other to new and more powerful births' (1967a: 81, 33). The Apollinian stands for clarity, lucidity, reason and rationality. It espouses controlled form with the 'urge' to make 'unambiguous' allowing artistic freedom only within a given 'law' (Nietzsche, 1968: 539). Apollinian artistic work demonstrates 'measured restraint' and 'freedom from the wider emotions' (Nietzsche, 1967a: 35–8).

In contrast, Nietzsche's Dionysian presents a shapeshifting and transgressive impulse which revels in ambiguity, immediacy, excess, sensuality, barbarity and the irrational; 'a passionate-painful overflowing into darker, fuller, more floating states' (Nietzsche, 1968: 539). These traits are expounded through the 'Dionysian content' of an artistic work (Nietzsche, 1967a: 37–54). Intoxication, and its phantasmal effect, which works on the imagination to produce a perception of 'intoxicated reality' (35–38), is paradigmatic of the Dionysian artistic impulse. The Dionysian impulse directly connects an individual with primordial, instinctive processes of perception and analysis, which is a crucial factor in the (syn)aesthetic appreciation process.

A slippage between states is integral to the metamorphic Dionysian impulse. In particular, any distinction between masculine and feminine is broken down in favour of primordial instinct in an embodied response. The Dionysian body is of consequence to the (syn)aesthetic style in that it is the primary receptacle for the intuitive processes of appreciation and the producer of philosophical analysis.[1] The Dionysian body provides a paradoxical sensuality, 'full of wisdom – a plurality with one sense' (Nietzsche,

1967b: 90). Such sensual (re)cognition enables a reconciliation of body and mind and can achieve a slippage between the noetic and chthonic. With Nietzsche's Dionysian impulse there is a fused possession of body and mind within artistic appreciation, working on the senses with an urgency that activates corporeal memory, imagination and intellect.[2]

Nietzsche argues for artistic work that enables an audience to 'feel most assuredly by means of intuition' (1967a: 46–8). In this way, the Dionysian impulse embraces immediacy in response and demands a complete perception via a visceral cognition. This substantiates a making-sense/*sense*-making faculty of appreciation which ensures that intangible ideas, states and experiences are made tangible, allowing a noetic 'sensing beyond' (after Nietzsche, 1967a: 132). Dionysian artistic works invigorate the 'aesthetic state', a primal and sensate mode of appreciation which draws on the (re)cognitive powers of the human body, a *'special memory*... a distant and transitory world of sensations' (Nietzsche, 1968: 422–7, emphasis added). Such intensity of perception produces a *'resonance'* where the spectator 'remembers and becomes aware of similar states and their origin' (Nietzsche, 1994: 22, emphasis original). Dionysian resonance is attributable to the corporeal memory activated within (syn)aesthetic perception and Nietzsche's aesthetic state comes close to the ineffable quality of the (syn)aesthetic-sense with its slippage between the noetic and chthonic.

Nietzsche's Dionysian impulse helps elucidate the creative approach taken within the (syn)aesthetic style as illustrated by all of the work discussed in Part 2. (Syn)aesthetic work is Dionysian dominant in impulse yet the controlled form of the Apollinian that harnesses the work prevents it from toppling into sensual indulgence, without form or considered content. This restraint also allows for the reflective contemplation, often subsequent to the work, which is influenced by the Dionysian intoxication of the senses. In this way the two combined activate a Nietzschean resonance extended by the Dionysian aesthetic state, which allows a 'sensing beyond' (after Nietzsche, 1967a: 132). These aspects endorse arguments for the (syn)aesthetic-sense.

(Syn)aesthetic work itself requires that the audience's interpretative faculties be influenced by an intoxication of the sensate body.[3] This fusion of mind and body to allow complete appreciation fully supports the (syn)aesthetic mode of visceral appreciation and combined somatic/semantic analysis. Also important to the (syn)aesthetic mode of performance and appreciation is the activation of the imagination which allows a (re)cognition of intangible states. The Dionysian impulse clarifies the immediacy of the innate response integral to the (syn)aesthetic style, where the audience do indeed experience and respond to the work in an immediate, intuitive and individual way.

Following this, an important feature of Nietzsche's theorization, which is fundamental to the (syn)aesthetic mode of appreciation, is the emphasis on immediacy in the Dionysian mode of aesthetic analysis. The Dionysian impulse heralds a theoretical approach which takes into consideration the immediacy of the disruptive experience of subconscious, corporeal and transgressive signifying practices.[4] (Syn)aesthetic interpretation denies a single, 'accepted' valuation, instead favouring individual reaction; a rejection of any single philosophical authority in favour of an individual's intuitive and creative response in the production and appreciation of any artistic work. (Syn)aesthetics is a Dionysian mode of practice and analysis as it prioritizes sensual perception and imagination yet engages cerebral powers of cognizance, measured Apollinian reflection, in an embodied way.

The Russian Formalists – Dionysian disruptions in linguistic play

> [T]he analysis of form understood as content.
> (Eichenbaum, 1965: 113)

> [T]o clarify the unknown by means of the known.
> (Shklovsky, 1965: 6)

> The sensitive ear will always catch even the most distant echoes of a carnival sense of the world.
> (Bakhtin, 1984: 107)

The Russian Formalists' concern with the form of verbal language is useful in clarifying the (syn)aesthetic style in relation to its disturbatory visceral-verbal *play*texts. Viktor Shklovsky argued for the act of *ostranenie* (literally, 'making strange'), more commonly understood as defamiliarization, as being integral to the production and reception of creative texts.[5] Shklovsky refers to *ostranenie* as 'roughened form' where defamiliarization 'results from special artistic techniques' that force the receiver 'to *experience the form*' (Eichenbaum, 1965: 113–14, emphasis added). Form is made unusual and therefore slows down perception ensuring that the experience of written or spoken language is perceived lucidly. It is this that allows a (re)cognition of language alongside the content it contains (situation, event, experience, emotion, and so on), enabling the audience to perceive anew.

Shklovsky argues that 'the acoustical, articulatory, or semantic aspects' of verbal language '*may be felt*' as a 'perceptible structure designed to be experienced within its very own fabric' (qtd. in Eichenbaum, 1965: 114, emphasis added). The purpose of such play with verbal language is 'to impart the sensation of things as they are perceived and not as they are known' (Shklovsky, 1965: 12). Osip Brik added to Shklovsky's ideas with his proposition that rhythm is no longer an abstraction but 'relevant to the very linguistic fabric' of the verbal play (Eichenbaum, 1965: 124). Rather than a 'superficial appendage, something floating on the surface of speech' (124) rhythm becomes an integral part of defamiliarized, sensate expression. Defamiliarized form ensures an experiential mode of interpretation via a 'special perception' where the cognizance of form transfers into this 'sphere of a new perception' via a 'unique semantic modification' (Shklovsky, 1965: 18, 21).

Of great importance to the (syn)aesthetic performance style is Mikhail Bakhtin's notions of 'carnivalization', a term that defines the shapeshifting effect that the substance of 'carnival' can have on artistic works (see Bakhtin, 1984). Of particular relevance to the (syn)aesthetic style is the association of carnival with sensuous experience and disturbing play. The associations with carnival, an inherently Dionysian act with its 'roots in the primordial order and primordial thinking', are the impulse behind 'carnivalized' works (107–22). Here the 'essence of carnival' which subverts everyday order, and is 'vividly felt by all its participants' demands 'ever changing, playful, undefined forms' (Bakhtin, 2001: 217–19). Regarding my focus on notions of slippage, Bakhtin argues '[c]arnival celebrates the shift itself, the very process of replaceability' and 'proclaims the joyful relativity of everything' (1984: 125).[6]

Bakhtin stressed that carnivalized practice can be used to 'disrupt authority and liberate alternative voices' that is those voices of difference assigned to the margins (Selden, Widdowson and Brooker, 1997: 42). Bakhtin's carnivalization provides the inception of 'polyphonic' works of art, which allows 'the development of a plurality of consciousnesses and their worlds' (Bakhtin, 1984: 17–18) so that 'voices are set free to speak subversively or shockingly' (Selden, Widdowson and Brooker, 1997: 44) without the author coming between individual 'consciousnesses' and the audience. In terms of the (syn)aesthetic appreciation process, carnivalization subverts textual practice in form and content and overwhelms traditional theoretical discourses via 'an indeterminacy, a certain semantic open-endedness' which produces 'counter-identification' or 'disidentification' (Bakhtin qtd. in Selden,

Widdowson and Brooker, 1997: 193) allowing the immediacy of individual interpretation to validate its own discourse. This also ensures that linguistic signification is defamiliarized, engaging alternative capacities for (re)cognition.[7]

The theories of the Formalists directly equate with the (syn)aesthetic approach to the written and verbal texts which enable an audience to interpret with a special perception due to the disturbance of language from its usual context. This unsettling of cognition thereby imparts the sensation of the object, state, experience or idea as perceived and not as it is known. Such defamiliarized linguistic play may trigger the (syn)aesthetic-sense. In this way, the audience experiences the sensate form of the language, which both heightens and dislodges its usual semantic making-sense procedure allowing for a somatic *sense*-making process to be of equal, or greater, significance.

Barthes – *jouissance* and pleasurable texts

> Does the text have human form, is it a figure, an anagram of the body? Yes, but of the erotic body.
>
> (Barthes, 1975: 17)

There is much in Barthes theorizing to support the potential that linguistic acts have in causing a (syn)aesthetic response via sensual and transgressive communication in the substance of the text itself. This results from a distinct (syn)aesthetic impulse in the original creation of the text. (Syn)aesthetic *play*texts are commensurable with Barthes' 'pleasurable text' which is a writerly 'text that discomforts' (Barthes, 1975: 14). In pleasurable texts a *'disfiguration* of the language' occurs to bring the receiver 'to a crisis' in their 'relation with language' (14–37, emphasis original). Barthes asserts that such text, as well as unsettling and disturbing an audience due to its form, provides *jouissance* (the closest translation being 'unspeakable bliss') where linguistic play 'granulates', 'crackles', 'caresses', 'grates', 'cuts', and 'comes' (67).[8] Jouissance names a fused physiological and psychological experience, simultaneously pleasurable and disturbing, that accentuates the human experience of the live(d), present (as in prae-sens) moment. It takes the actual body beyond that which is already known, arguably into the realms of Kant's negative pleasure, as discussed in Chapter 1 above. The disturbance of sensations resulting from the defamiliarized state of jouissance produces a corporeal and cerebral impact which engages a (syn)aesthetic-sense and causes a reperception allowing an individual to experience anew, consider anew, interpret anew.

Barthes' thinking is useful in elucidating visceral-verbal *play*texts as he does not confine his theories to the written text, as much linguistic analysis does, but to that which is performed; *'writing aloud'*, or 'vocal writing' (66, emphasis original). Writing aloud is an entirely corporeal act that foregrounds, 'the articulation of the body', a physicality of verbal play that conveys 'language lined with flesh' (66). Here semantic meaning is secondary to the immediate experience of the sensate nature of verbal play as a corporeal linguistic delivery shapeshifts 'the signified a great distance' (67). This causes a (re)cognition of language which allows the audience to experience the work physiologically. Following this, it is an 'anterior immediacy' (Barthes, 1982a: 439) in the corporeal memory of the (syn)aesthetic experience of these words in performance, that instils the prae-sens of that moment of experience in subsequent interpretation.

Barthes thinking validates (syn)aesthetics as an interpretative device by demanding modes of interpretation and theorizing that belong to the immediate and the innate. That is, analytical discourse that meets the creative impulse of the texts under scrutiny and articulates the slippage between the corporeal, transgressive and ambiguous in vocabulary, style and application. Such analysis acknowledges the sensual impact on the cerebral and celebrates the body as the sensate conduit of the overall experience of the work. This enables 'meaning' to be sensually, and individually, produced and theorized in a way that matches the corporeal, pleasurable nature of the work itself.[9]

Barthes' pleasurable text aids understanding of the (syn)aesthetic impulse and response in performance as he argues for writerly texts that are created, delivered and appreciated corporeally. Such writing clarifies the style of (syn)aesthetic *play*texts in that both form and delivery of the text cause the receiver to reperceive and recognize the embodied nature of verbal language in an experiential way.

Kristeva's *semiotic chora* and genotext

> The transfinite in language, as what is 'beyond the sentence', is probably foremost a going through and beyond the naming. This means that it is going through and beyond the sign, the phrase, and linguistic finitude.
>
> (Kristeva, 1992f: 190)

Kristeva provides an argument for a primordial site of communication which supports the fusion of the verbal and physical as an important means of signification. Her arguments for a *semiotic chora* ('chora' from the Greek 'distinctive mark, trace, precursory sign, imprint, figuration') and 'genotext' help to clarify the (syn)aesthetic modes of communication within performance (see Kristeva, 1999a).

The *semiotic* and *symbolic*, similar to the Dionysian and Apollinian, are 'two modalities...inseparable within the signifying process that constitutes language' (92). The symbolic is the modality, associated with reason, repression and control, that produces syntax, fixing form in the process of linguistic making meaning. Kristeva's semiotic, in contrast, is useful to understanding (syn)aesthetics in performance and appreciation as it presents a chthonic site of articulation which expresses innate and primary processes that are 'pre-sign, pre-meaning' (Kristeva, 1982: 212, n.3) able to tap a pre-verbal consciousness. It provides a discourse for the unconscious and the body – both internal and external human experience through a 'recasting of language' (61). Kristeva's semiotic recognizes the significance of the visceral and the inarticulable in affecting individuals in a meaning making way. Here analysis and signification 'refers back to an instinctual body' that 'ciphers the language' (Kristeva, 1992a: 146), ensuring that, 'sense topples over into the senses' (Kristeva, 1982: 140).

The chora is 'the dimension beneath the surface of signification' (Kristeva, 2000: 268), a space that articulates passions and drives, demanding liberated structures for expression and interpretation. It is associated with unconstrained connections and transgression. Here an individual's conscious and unconscious impulse, and varied processes of interpretation and 'meanings' reached, are simultaneously connected, allowing for an 'intertextuality' that is 'translinguistic' (Kristeva, 1992e: 36). The semiotic chora expounds a chthonian rhythm, that retains a Dionysian '*repetition* and *eternity*' (Kristeva, 1999c: 191, emphasis original), imposing a primordial temporality which may 'shock' yet also be experienced as an 'unnameable jouissance' (191).[10] Such rhythms and repetition dominate in (syn)aesthetic performance texts.

Following this, for Kristeva, any text that is close to, and includes, semiotic processes is a 'genotext' (1999a: 120).[11] Defined as 'language's underlying foundation' the genotext is 'not linguistic' but a 'process' (121), which construes a shapeshift discourse (creative and theoretical) and explores transgression and corporeality in all signifying practices. This articulates the fused textual practice of the (syn)aesthetic style.

Kristeva highlights how the persistent influx of the semiotic destroys and reinvents the symbolic order. This constant 'tearing open' of the symbolic by the semiotic allows the 'interplay of meaning and jouissance'

(Kristeva, 1992: 148) and necessitates a 'transgression' that enforces the shifting signifying practice 'called "creation"' (Kristeva, 1999a: 113). This transposition from the semiotic to the symbolic holds a key to understanding the flux in making-sense/*sense*-making. Within this flux, Kristeva argues that certain artistic works have the potential to 'reach the semiotic chora' and 'destroy the symbolic' (see Kristeva, 1999a: 103–22).

A (syn)aesthetic performance engages the semiotic chora primarily in its fusion of conscious and unconscious perception. The signifying processes of the chora is also activated when transgressive signifiers come into play in (syn)aesthetic work, particularly when the conceptual is explored through the body and the corporeal is explored through speech as occurs in the work of Churchill, Kane or Wallace. The idea of the performer as a site, sight and cite for the exploration of verbal and physical, internal and external, and cerebral and corporeal experience in (syn)aesthetic work evidences Kristeva's notion of a human subject as an intertextual 'play of signs' (qtd. in Broadhurst, 1999a: 6).

The semiotic chora, the genotext and their inherent chthonic processes of communication support and clarify the fused communication and perception capable in (syn)aesthetic performance. Kristeva's semiotic recognizes the significance of the sensate and visceral in affecting individuals in a double-edged making-sense/*sense*-making way. Such linguistic analysis, responsible for reclaiming the spoken and written as a physical and physiological act supports the (syn)aesthetic style's somatic/semantic function and its quintessential visceral-verbal feature.

Cixous and Irigaray – *écriture féminine*

> The essence of nature is now to be expressed symbolically ... and the entire symbolism of the body is called into play.
> (Nietzsche, 1967a: 40)

> Write yourself: your body must make itself heard.
> (Cixous, 1993: 97)

> We have to discover a language which does not replace the bodily encounter ... but which can go along with it, words which do not bar the corporeal, but which speak the corporeal.
> (Irigaray, 1985: 43)

Écriture féminine, Dionysian in impulse, is both a sensate writing practice and an analytical tool. As a fused creative and critical methodology it establishes a multiple perspectival process 'another way of knowing. ...Another way of producing...where each one is always far more than one' (Cixous, 1993: 96), highlighting a fusion of ambiguity, slippage and transgression.

Écriture féminine, like Kristeva's semiotized genotext, establishes a Nietzschean, shapeshifting eternal recurrence as it is a 'feminine morphology' (Irigaray, 1999e: 55) which is always in the process of becoming. As Irigaray asserts, *'form is never complete in her'* and it is this 'incompleteness' of form that allows the 'feminine', as a writing effect, to eternally morph its own morphology, becoming, 'something else at any moment' (55, emphasis original). This eternal recurrence is also evidenced in the fact that codes and conventions of creative practice must be continuously overturned and replaced by *'a new insurgent writing'*, a writing of the body where 'the huge resources of the unconscious...burst out' (Cixous, 1993: 97, emphasis original).

Écriture féminine establishes a linguistic practice 'which does not replace the bodily encounter' but goes along with it (Irigaray, 1999a: 43) paralleling Barthes' pleasurable text and Kristeva's semiotic chora.[12] The feminine, 'does not deny unconscious drives the unmanageable part they play in speech' (Cixous, 1993: 92). Furthermore, Irigaray argues that the Dionysian impulse of écriture féminine 'intervenes between body and soul...endlessly pulling down the barrier between them' (1991: 129). This dualistic intervention makes the process 'accessible to the senses' (134). In embracing the corporeal, écriture féminine prioritizes sensate communication and sensate perception, 'does not privilege sight' but 'takes each figure back to its source, which is among other things, *tactile*' (Irigaray, 1999b: 126–7, emphasis original). In prioritizing the body as the source and morphology of creative practice, écriture féminine supports the emphasis on the body as producer and receiver of signification in (syn)aesthetic performance work.

Écriture féminine aims to break down any rigid oppositions (especially of feminine/masculine) to embrace and take pleasure in the slippage in-between, which for Cixous, enables a return to the primordial. Cixous re-evaluates this Dionysian plurality and blurring of opposites as 'the location within oneself of the presence of both sexes...the non-exclusion of difference or of a sex' and asserts that there is a need for the 'I/play of bisexuality' (Cixous, 1993: 84) in order to create work that affects in a visceral way. It is in performance that écriture féminine is most present (as in prae-sens) providing a creative site where 'it is possible to get across the living, breathing, speaking body' (Cixous, 1995a: 134).

Écriture féminine, an inherently Dionysian discourse, connects with Kristeva's semiotic chora, celebrating the slippage between the conscious and unconscious. It highlights sensate access and sensate pleasure as the ultimate form of artistic appreciation. Écriture féminine is particularly useful to the (syn)aesthetic style in that it demands sensual approaches to writing text and in doing so it establishes a creative methodology and discursive practice that can also be applied to *the body as text* in performance. In this way, Cixous and Irigaray provide a critical discourse for Artaud's 'writing *of* the body' (Derrida, 1978: 191, emphasis original), fusing the physical with the verbal, and analysis with practice.

Artaud – disturbance and sensation in the Theatre of Cruelty

> One cannot separate body from mind, nor the senses from the intellect, particularly in a field where the unendingly repeated jading of our organs calls for sudden shocks to revive our understanding.
>
> (Artaud, 1993: 66)

Artaud's manifesto for a Theatre of Cruelty underpin the (syn)aesthetic style and its experiential strategy of appreciation. Artaud's Cruelty asserts performance techniques that aim to (re)connect body and mind through the prioritization of the human body within performance. Artaud demands a 'total' experience for performer and audience alike through a manipulation of all the elements of theatre via unusual and exciting gesamtkunstwerke. Artaud demanded a new taxonomy to scrutinize and understand such performance, where the work itself 'governs' the 'commentary' applied (Derrida, 1978: 175).

Artaud's insistence on the actual body as the primary performance signifier presents 'another form of writing...the writing *of* the body itself' (191, emphasis original). It is this commitment to the body as the primary signifier and translator of a visceral performance language that foreshadows a physical practice of écriture féminine, employing the actual body as the performing and receiving receptacle to express and comprehend chthonic states. It thereby defines a mode of performance practice that signifies through the semiotic chora.[13]

A crucial aspect of Artaud's theatre which supports the (syn)aesthetic style is the power to disturb and enliven through the interaction of the live physical body in performance affecting the sensate physical body in the audience. As Nigel Ward posits, in Artaudian practice it is 'the body of the actor' that works 'directly upon the nervous system of the audience' (1999: 124). The cerebral appreciation of the work is part of this fused experience as it is connected with the body, allowing semantic and somatic readings to combine. In Artaud's theatre this is a direct result of the fact that 'the brain is another organ to be acted upon...in the same way as the rest of the body' (124). Artaud states that 'cruel' theatre 'upsets our sensual tranquillity' and 'releases our repressed subconscious' (1993: 19). This ensures that those inner, primordial states and experiences which are considered inexpressible are (re)cognized cerebrally as a direct result of the corporeal 'communion' between performer and audience, where the inexpressible is 'made to enter the mind through the body' (1993: 77).

Artaud's theories articulate the somatic/semantic nature of the (syn) aesthetic style where the sentient source and conduit of the body can affect a (syn)aesthetic-sense which allows a visceral cognition of intangible states. This is equivalent to Artaud's *'active metaphysics'* where the performance mode affects 'on all conscious levels and in all senses' which leads to 'thought adopting deep attitudes' (33, emphasis original).

Artaud's arguments regard performance as 'an immanent, damaging, purging event' (Ward, 1999: 123) that 'wakes up...heart and nerves' (Artaud, 1993: 64). This transforms the mind/body connection via a 'tangible laceration' inflicted on 'the senses' (Artaud, 1993: 65). Heralding a (syn)aesthetic style of production, such a performance mode is 'enacted upon the body, to assault the senses' (Ward, 1999: 123) in an entirely Dionysian manner. The aim is for the audience to experience a 'metamorphosis' within this 'total experience' through the disturbatory impact of these 'visceral assaults' (128). Such a visceral experience produces a fused 'consciousness' of 'exposed lucidity' (Derrida, 1978: 242) where body and mind, senses and intellect are conjoined through the 'sudden shocks' which revive understanding in a total and sensate way (see Artaud, 1993: 66). The experiential quality of performance defined by Artaud affects a practice of jouissance and foregrounds the visceral impact that the (syn)aesthetic style has on the individual's body in appreciating the work.

Artaud's theories underpin the (syn)aesthetic performance style, in form and content, as it embraces a hybrid mode and foregrounds the body in performance as sight, site and cite of disturbance and jouissance. Furthermore, Artaud's principles of taxonomy, where the work itself and subsequent appreciation govern the analysis, affirms the need for a (syn)aesthetic contact in performance analysis that engages a making-sense/sense-making faculty in the processes of appreciation.

Novarina – corporeality and carnage in the Theatre of the Ears

> We write in confrontation, through the love of hand-to-hand fighting with our language. ... My language is not in my mind, like a tool that I would borrow in order to think. It is entirely within me: words are our true flesh.
>
> (Novarina, 1996: 125)

As a theatre practitioner and theoretician, Novarina puts forward a strong argument for the visceral impact of the writerly and verbal act in performance. His theories accentuate the corporeal, especially the physiological, sexual and gastronomical, aspect integral to the creation, rehearsal, performance and interpretation of (syn)aesthetic *play*texts.

Novarina talks of speech as 'the *speak*' that is 'most physical in the theater' (1996: 58, emphasis original), echoing Barthes' theories of the jouissance of 'vocal writing' (Barthes, 1975: 66). Novarina demands liberation of the body into verbal acts in order to overturn the Cartesian implication that 'words fall into our heads from the heavens, that is thoughts which are expressed, and not bodies' (Novarina, 1993: 100). He talks of 'articulatory cruelty, linguistic carnage' within the creation of the text suggesting that via such brutal and disturbatory manipulation of verbal language 'perceptions' can be 'changed' (96–9).

Novarina parallels Artaud's call for a holistic performer, body and speech fused in the processes of production, where the 'entire body must come into play' (Novarina, 1996: 108). The performer must 'put his body to work... sniffing, chewing, breathing in the text... vigorously working it over' in order to 'discover how it breathes and how it is rhymed' (Novarina, 1993: 101). Such a physical wrestling with the text enables 'a profound reading, ever deeper, ever closer to the core' (101). In this way Novarina asserts that 'the text becomes the actor's nourishment, his body' and he echoes Artaud and Cixous in his demands that performers, 'rewrite' text with the 'body' (101) for 'words are our true flesh' (Novarina, 1996: 125). Novarina asserts that, 'the text is nothing but footprints on the ground left by a dancer who has disappeared' and as a result the performance of the text is 'a matter of manifesting, of soliciting, the existence of something that wants to dance' (1993: 102).

Allen S. Weiss posits that Novarina follows Artaud in establishing 'a theatrical practice that leads well beyond the textual, directly into the morass of the body' (Weiss, 1993: 85). Novarina demands an appreciative mode which allows the body to 'open up' its 'mental flesh'

(Novarina, 1996: 64) and experience the work produced via an 'amorous interchange' (108) between text and performer, performance of the text and audience. This enables a 'reconciliation of word and body' (Weiss, 1993: 86) in interpretation.

It is the case that 'the refusal of meaning and the reduction of speech to the pure voice, of language to the body' (88) in Novarina's theories can be a significant feature of the (syn)aesthetically styled performance and *play*text. Verbal language which plays with 'levels, and not origins, of meaning' (88), shifting the traces of semantic meaning integral to words, encourages an audience to interpret meaning on both a semantic level that is present, if somewhat disfigured, and also to (re)cognize this meaning within the simultaneous reperception of the *sound* of the word. Here 'the phonetic elements of words', concerned with 'meaning and usage', are ignored in favour of a somatic response where the meaning of words is reflected in the sound they embody (Luria, 1969: 86). In this way 'perpetual linguistic shifts and stresses' force a reconstitution of an individual's 'lexicon' and 'thought' (Weiss, 1993: 92). Novarina's theory and practice, like Artaud's, play within Kristeva's semiotic chora where 'texts composed of glossolalia which mean nothing and are totally explosive' become 'no longer language but pure drive' (Kristeva, 2000: 265).

Novarina meets Barthes, Cixous and Irigaray by establishing a performance writerly practice that returns to the corporeal in the *act* of writing and the subsequent translation to performance. His theories are important to the (syn)aesthetic performance style as he highlights the corporeality of verbal language within the full process from creation through production to appreciation. He stresses how the visceral brutality of such defamiliarized and carnivalized texts allows a reperception of language and meaning to occur; a reperception that invigorates the imagination and is primarily interpreted through the body in a sensate manner.

Fundamental to my argument for (syn)aesthetics is Novarina's argument for the act of writing as a physical performance practice itself which collaborates with the processes of performance, highlighting the exchange of corporeality between writer and performer, 'to change bodies...to breathe within another's body' (Novarina, 1996: 108), emphasizing how the traces of the living body that writes remain within the fibres of the text to be interpreted by the living body that performs. This foregrounds the corporeal 'sharability of sentience' (Scarry, 1985: 326) between all participants, including the audience, within such creative practice.

With his consideration of sound and 'verbigerations' (Weiss, 1993: 84), Novarina answers Artaud's demand for organic human sound in performance, alongside the manipulation of spoken language 'in a new, exceptional and unusual way, to give it its full, physical shock potential' (Artaud, 1993: 35). Novarina provides a bridge between Artaud's physical theatre and Barker's writerly theatre, confirming the transgressive nature of (syn)aesthetic practice as he asserts the potential of primitive sounds as much as the eloquence of visceral, verbal language within writerly *play*texts to communicate via Kristeva's semiotic dimension.

Barker – imagination and disturbance in the Theatre of Catastrophe

> In the anguished, catastrophic times we live in, we feel an urgent need for theatre that ... arouses deep echoes within us.
> (Artaud, 1993: 64)

> This pain is necessity. The Theatre of Catastrophe is not the comfort of a cruel world, but the cruelty of the world made manifest and found to be – beautiful.
> (Barker, 1997: 116)

> To speak is truly catastrophic.
> (Novarina, 1996: 134)

Barker's arguments for a Theatre of Catastrophe are useful in elucidating the (syn)aesthetic style as he establishes a writerly performance practice that reve(a)ls in the sensate quality of verbal language and emphasizes the importance of a disturbed imagination within audience appreciation. Barker's notion of language as a physical act that is both sensual and cruel immediately draws on the dualistic Dionysian impulse. Echoing Artaud and Novarina in his assertions for speech as a sensual act, Barker asserts that the transgressive and disturbatory potential of the verbal act, 'breaks the bonds of the real, disrupts the familiar' thereby instilling a Dionysian 'intoxication' in appreciation which 'subverts reason' (Barker, 1997: 213).

Barker argues for the need to return an audience to the primordial imagination where, once engaged the audience will no longer seek coherence but instead experience 'moment by moment' and 'contradiction by contradiction' (38). The 'imagination' here then expresses both the human mental, visual-imaginative capabilities alongside the noetic 'secret' (166) potential of human perception. Imagination articulates and finds form for that which is intangible and helps stimulate the emergence of a (syn)aesthetic-sense. Here Barker's writerly practice links closely with the highly physical signification in Artaud's Cruelty via the re-examination of 'all aspects of the inner world' in order to position 'imagination's rights in the theatre once more' (Artaud, 1993: 71). Barker's insistence on the imagination in performance establishes work that is 'tentative, speculative' and creates 'anxiety in the beholder' (Barker, 1997: 135). Here Barker echoes Kant's negative pleasure, where the experience of pleasure 'is only possible through the mediation of a displeasure' and such a process 'strains the imagination to its utmost' (Kant, 1911: 91–120).

Barker's theories articulate the need for a performance style which works on the imagination as well as the senses and returns performance to an 'other-worldliness' (Barker, 2001: 2). His arguments immediately correlate with the (syn)aesthetic style which highlights a 'primitive sensitivity' and breaks down the boundary between the 'real' and the 'imaginable' (Luria, 1969: 80, 144). This vivid and lucid imaginative capacity disturbs in order to (re)awaken ideas and experiences within its audience, liberating the imagination and powers of cognizance within this disturbance, causing an 'exposed lucidity' (Derrida, 1978: 242). Barker's arguments insist on the audience drawing on their own experience within, and beyond, the performance moment in order to find 'meanings' for themselves where 'powers of reconciliation or resolution are abolished in favour of a passionate assertion of human complexity' (Barker, 1997: 79).

Barker asserts that it is the job of the audience to work through the difficulties of interpretation, work out the difficult form and content through complete, corporeal cogitation. In this way individuals within the audience have the 'rights of interpretation' which causes a creative tension between the 'audience and the stage itself' (51–2). Performance, from conception to reception, is a journey of unknowing for writer, director, actor and audience alike which 'insists on the limits of tolerance' and 'inhabits the area of maximum risk, both to the imagination and invention of its author, and to the comfort of its audience' resulting in meaning being derived 'from the dissolution of coherent meaning' (52–3).

Barker's theories support and elucidate the (syn)aesthetic style as he affirms the need for a visceral-verbal writerly practice that disturbs via its challenging form and in doing so, attempts to articulate that which is hidden, or intangible. The insistence on the fusion of the sensual, beautiful and cruel within writerly practice foregrounds the visceral nature of the (syn)aesthetic playtext. Barker further substantiates the (syn)aesthetic mode of appreciation in his insistence on an individual's 'rights of interpretation' (51). In (syn)aesthetic analysis an innate, individual interpretation is prioritized in performance work that has been experienced 'moment by moment' (38). Also important in clarifying the form and content of (syn)aesthetic performance is the transgressive nature of Catastrophe where 'theatre is law-breaking' (Barker, 2001: 3) in order to disturb, destroy and (re)create notions of production, appreciation and interpretation.

Broadhurst and the liminal – touching the edge of the possible

> All liminal works confront, offend or unsettle.
>
> (Broadhurst, 1999a: 168)

> In digital practices, due to the hybridization of the performances...various intensities are at play. It is these imperceptible intensities, together with their ontological status, that give rise to new modes of perception and consciousness.
>
> (Broadhurst, 2007: 5)

Broadhurst's theory of liminal performance follows Victor Turner's arguments for the liminal as a Dionysian site of 'fructile chaos, a fertile nothingness, a storehouse of possibilities, not...a random assemblage but a striving after new forms and structure', placing 'greater emphasis on the corporeal, technological and chthonic' (Broadhurst, 1999a: 12). Liminal performance can be described as being located at the 'edge of the possible' (1). Its quintessential aesthetic features are hybridization, indeterminacy and the collapse of the hierarchical distinction between high and popular culture. The quasi-generic traits of liminal performance are experimentation, heterogeneity, innovation, marginality, 'a pursuit of the almost chthonic' and an emphasis on the 'intersemiotic' (12–13).

Following Derridean 'wide, jarring metaphors', utilizing the latest developments in media, digital and biotechnologies (see Broadhurst, 1999a: 10–13; 2007), the liminal aesthetic highlights the visceral and transgressive strategies of the (syn)aesthetic style. Derridean jarring metaphors 'unsettle the audience by frustrating their expectations of any simple interpretation' (175). They also demonstrate a 'refusal to prioritize any of the senses (visual, aural, tactile, olfactory)' in favour of emphasizing 'their interconnectedness' (Broadhurst, 2007: 30). These jarring metaphors are developed in a concrete manner in Broadhurst's attention to liminal hybridity which helps to clarify the (syn)aesthetic hybrid.

Hybridized performance 'simultaneously distances and engages the spectator' and establishes form as 'a merging of the aesthetic with everyday life', 'montage', 'dreamscape', 'collage' and 'imagination', thereby instilling 'lasting effects' (Broadhurst, 1999a: 71–89). In answer to the assumption that all performances present a gesamtkunstwerk, as discussed in more detail in Chapter 3, Broadhurst clarifies that hybrid performance should be taken to mean that which combines disparate disciplines *in order to undermine* 'accepted boundaries and definitions' (1999b: 24).

A significant aspect of liminal performance is that it continuously challenges traditional aesthetic concepts due to its indeterminate nature of process and production. As a result, inherent experiences from the audience are responses of disquiet and discomfort. Features such as 'immediacy, disruption and excess' (Broadhurst, 1999a: 171) are Dionysian dominant traits inherent in liminal performance, which presupposes the Dionysian impulse at the root of liminal work. Liminal performance relies on the corporeal 'transmission of primarily emotive experience' (79). This, as Broadhurst points out, corresponds to the Dionysian impulse in appreciation of 'tremendous awe which seizes man when he suddenly begins to doubt the cognitive modes of experience' (Nietzsche qtd. in Broadhurst, 1999a: 105). Liminal performance seeks to bring about a 'consciousness' via 'emotive experience' which produces an opposition to traditional, mainstream performance modes 'on the basis of...the disruption of the emotions' (79). This impact of corporeal and cerebral disturbance in the reception of liminal work immediately suggests a shifting between sense/*sense* in interpretation which demands a (syn)aesthetic process of analysis.

Just as Barker foregrounds the 'discomfort' and the 'irrational', Broadhurst highlights the disturbing nature of the liminal in that, 'all liminal works confront, offend or unsettle' and leave 'many spectators exhausted by the end...overwhelmed by the emotional complexity of the experience' (168, 71). In this way it is the experience of sensual disturbance, as a result of a transgressive hybridized mode which takes the audience 'to the edge of the possible' (1).

In liminal performance there is an insistence on the audience drawing on their own experience within, and beyond, the performance moment in order to find 'meanings' for themselves. Broadhurst argues that a 'lack of resolution or closure is a central trait of liminal performance' (71). Within liminal theatre, the 'free association of themes rather than a linear narrative' subverts logical explanation in favour of an immediate and innate response where a spectator 'is required to turn to his or her own life experiences' (77).

The liminal is useful in clarifying the (syn)aesthetic hybrid and its potential for visceral disturbance caused by an individual's process of becoming aware of the special fusion of diverse performance languages. With its hybridization and emphasis on the intersemiotic; its pursuit of the chthonic; its foregrounding of the actual body and its use of Derrida's 'wide jarring metaphors' (10), the liminal clarifies the features present in the (syn)aesthetic style and prioritizes the need for an intersemiotic approach in performance analysis that is substantiated in (syn)aesthetic strategies of analysis and interpretation.

Broadhurst foreshadows the (syn)aesthetic mode of appreciation in that liminal performance demands that 'neither the ingredients to be judged nor the toolkits of analysis are given' but instead 'elaborate one another in a progressive dynamic' (19). She heralds the (syn)aesthetic mode of analysis in arguing that '[l]iminal performance demonstrates a need for a new form of aesthetic interpretation' that identifies the exciting and unsettling experience for the audience provoked by such work, allowing for 'intersemiotic modes of signification' (171–8).

Broadhurst's neuroaesthetic approach to digital practice in the arts explores new interpretative frameworks to articulate performance practice (see 2007). A key feature of neuroaesthetics is the identification that certain art forms, of which Broadhurst includes the liminal quality of digital practice, are able to 'tap into innate "form primitives" of the brain which are not as yet fully understood' (Ramachandran and Hirstein qtd. in Broadhurst, 2007: 59–60). Broadhurst's focus on neuroaesthetics anticipates the shifting and sensate central features of (syn)aesthetic appreciation. She calls for a new theorization, which incorporates neuroaesthetic approaches and adjusts critical theory to allow for 'technical interface' and 'corporeal prominence...however transitory or virtual' (16). Where Broadhurst turns to selective aesthetic approaches offered by Derrida, Gilles Deleuze and Félix Guattari, Jean-François Lyotard and Merleau-Ponty, arguably (syn)aesthetics is an interpretative strategy that is wholly 'capable of addressing the sensate disturbances evoked by much of the sophisticated technological art practices' that play with the actual and virtual corporealites, that Broadhurst scrutinizes (see 2007: 19).

(Syn)aesthetics – connecting theories

> The sensuous moment of knowing.
>
> (Taussig, 1993: 45)

These critical and performance theories are collected and summarized here to support and clarify the terms of (syn)aesthetics. Fused together they point toward the (syn)aesthetic approach to analysing and interpreting work. They also help to elucidate certain features of the artistic style. Important to each is the presence of intense sensual experience coupled with considered discipline in form which avoids any unnecessary over indulgence in sensuality 'for the sake of it'. As a result measured reflection that is underpinned by the sensual experience is demanded by the work in immediate and/or subsequent appreciation.

These embodied perspectives have also been connected in this chapter to foreground how, as a performance theory, (syn)aesthetics does not push the potential of linguistic practice to the background nor does it focus on verbal practice as the foremost language in performance. Instead it fuses these theories in order to articulate a response to work which plays with the possibilities of an interdisciplinary approach. (Syn)aesthetics prioritizes a *style* of practice which is attributable to certain writerly theatres just as it is present in multimedia events and purely physicalized performance.

The following chapter surveys these strategies in practice. In doing so it identifies how key ideas and terminology from Chapters 1 and 2 are made applicable in (syn)aesthetic analysis and shows how they operate in the visceral performing style that has previously proven a challenge to articulate due to its diverse and experiential nature.

1.3
(Syn)aesthetics in Practice

> Believing in the power of theatre is like believing in religion: you have to experience its effect in order to understand the attraction of it.
>
> (Richard Eyre and Nicholas Wright, 2000: 11)

Figure 1.3 Image from Shunt's *Dance Bear Dance* (2002–3), courtesy of Lizzie Clachan. Reproduced with kind permission of Shunt Theatre Collective. Photo: © Shunt

The experiential impact of (syn)aesthetic performance affects a cognition which leaves its traces on the perceiver's body via the imme-diacy of a corporeal memory. Put simply, we *feel* the performance in the moment and recall these feelings in subsequent interpretation. The nature of such appreciation engages a double-edged making-sense/*sense*-making process. This can affect an ineffable quality; a (syn)aesthetic-sense. The (syn)aesthetic style is a performance approach which, in form, content and methods of analysis emphasizes fused sensual experience. Although I detail three key strategies here, it is important to note that the consideration of the (syn)aesthetic hybrid, consisting of all manner of performance techniques and disciplines, highlights how (syn)aesthetically styled work results from varied artistic fusions within interdisciplinary practice. In (syn)aesthetic performance there is a constant slippage and exchange between the dominance of any one of the three key strategies surveyed here, as illustrated by all of the discussions in Part 2.

The (syn)aesthetic hybrid – a 'total' (syn)aesthetic

> The triumph of pure mise en scène.
> (Artaud qtd. in Derrida, 1978: 236)

When considering the (syn)aesthetic hybrid it is necessary to acknowledge the fact that any theatre work manipulates various design and performance techniques within the staging, which arguably renders the term hybrid unnecessary. However, with the (syn)aesthetic hybrid, the term is employed to refer to the particular way in which these elements are fused *in order to* generate a visceral quality within the processes of production and appreciation.[1] It is the distinctive nature of the exchange within the (syn)aesthetic hybrid that procures a defamiliarized mix of the aural, visual, olfactory, oral, haptic and tactile within performance, enabling a (re)cognition of form due to the unsettling and/or exhilarating *process of becoming aware* of this special fusion. With the (syn)aesthetic hybrid the combination of forms and techniques amounts to an additive experience in reception, which transgresses notions of what performance is and can be. With its slippage between disciplines, blending a variety of forms and techniques from high and low culture, the (syn)aesthetic hybrid is inherently Dionysian in impulse. This is clear in the shifting realms of Curious, Punchdrunk and Shunts' work through the interdisciplinary practice of Bodies in Flight, Carnesky, Graeae and Khan to the hybridized forms demanded by the writing of Churchill, Kane and Wallace.

A (syn)aesthetic hybrid embraces interdisciplinary practice from a variety of historical theatre and dance conventions to stand-up, stripping, puppetry, film, design, technology and more, playing with juxtaposition. It may fuse disciplines to the extent that demarcation is impossible, or alternatively, any one element can dominate at any time. In this way it develops The Russian Formalists' literary ideas in that it seeks to explore content through form in an original and unusual manner and consequently the very form of the performance takes on a polyphonic quality. The fused elements can provide a concrete haptic and tactile rendering of different consciousnesses and experiences with a carnivalized layering of the ritualistic, the sacred and the profane.

A (syn)aesthetic hybrid answers Artaud's demands for 'total theatre', combining speech, movement, dance, design, sound (organic and/or composed), light, puppetry, mask, technology and site to 'transgress the ordinary limits of art and words' (see Artaud, 1993: 68–87). Such a fusing of aesthetics ensures that the performance space speaks 'its own concrete language', a tangible 'many-hued spatial language' which 'develops all its physical and poetic effects on all conscious levels and in all senses' with exhilarating and disturbing results (27–45). Following Artaud, within a (syn)aesthetic hybrid there is a fusion of each visceral performance language to ensure that, 'connections, levels, are established between one means of expression and another' in order to 'fuse sight with sound, intellect with sensibility' (38–73). As Jones of Bodies in Flight asserts in Part 2:

> There are no hard connections between the different elements. It's more like resonances. The word we've been using a lot lately is 'mood' because it implies connections between things but doesn't say they translate.

In contemporary practice Artaud's theories are enhanced by the exciting advances in performance techniques, design including light, sound, film, digital technologies, on-line interaction and site. An exciting element of the (syn)aesthetic hybrid is the ongoing experimentation with site-specific/site-sympathetic work, as exemplified by the work of Carnesky, Curious, Punchdrunk and Shunt.[2] Site not only adds to the (syn)aesthetic interpretation of performance events but site-specific/site-sympathetic events also ensure that theatre is no longer placed in a darkened auditorium, behind heavy velvet curtains, but moves and breathes anywhere, the site itself inspiring and (shift)shaping the work.

Figure 1.4 Punchdrunk's *Sleep No More* (2003). Performer: Geir Hytten. Photo: Stephen Dobbie. Image © Punchdrunk

The architectural impact of site makes the audience aware of the haptic quality of spatial presence and their position within that. With site-specific performance the workings of the (syn)aesthetic hybrid ensure that space truly becomes 'a tangible, physical place' (27). This makes the theatrical experience multidimensional and produces textural layers of meaning for the audience to absorb and interpret. Form becomes a dynamic entity, crafted in such a way that the audience 'experiences' its sensual quality *actually* as a 'perceptible structure designed to be experienced within its very own fabric' (Shklovsky qtd. in Eichenbaum, 1965: 114).

In the (syn)aesthetic hybrid the design is woven into the fabric of the performance in order to create tangible sets and to suffuse the space with a palpable sensibility as with Punchdrunk's *Masque of the Red Death* (2007–2008) where the experience is of having taken laudanum with Edgar Allen Poe and physically entered his *Tales of Mystery and Imagination*. Likewise, Shunt's *Tropicana* (2004–2005) transports the audience to a disturbing underworld of rock-showgirl funerals and nightmarish scientific experimentation. Design elements become vital experiential aspects of such events, playful sites that offer a physical realm that is at once chthonic and noetic.

In site-based hybrids the audience is often required to become an affective element within the design itself. With Punchdrunk's work, the audience is individually masked for anonymity and invited to enter a sensual world in order to play within the space and the performance itself. The mask encourages risk taking, a stepping outside of the self.

Furthermore, the masked audience adds to the otherworldliness of the experience; the audience frame the sequences as they watch, often unwittingly, breathtakingly choreographing themselves into beautiful, carnivalesque sculptures. In this way the audience becomes part of the live performance, the design and the architecture. Shunt's unique approach to the manipulation of audience and space as an interactive prae-sens ensures that sensual perception and interpretation is paramount. For instance, in Shunt's *Ballad of Bobby François* (1999–2001) there is a multisensory destruction of an aeroplane fuselage, occurring in a blackout with an evocative sound score and the tangible sensation of the machine being destroyed around and within the audience. Audience imagination in the moment and any ensuing narrative and thematic impressions are triggered by this spatial and aural assault on the senses.

In terms of experiencing form in its very own fabric, in Part 2 both Punchdrunk and Shunt highlight the careful shaping of the audience as part of the hybridized form in order to enable the necessary quality of affective experience appropriate to each piece. As Doyle puts it, the audience 'become part of the choreographic landscape'. Similarly, Clachan of Shunt describes how the audience are designed to be fundamental to 'the rhythm of a performance'. Incorporating the audience as an architectural and choreographic element of the hybrid in this way establishes a unique and palpable rhythm within the form. This in turn transfers to the haptic rhythm experienced by each audience member in the moment and in subsequent recall.

In any (syn)aesthetic hybrid, whether site based or within traditional theatre environments the design is often integral to the imaginative play of the work as it is shifted within the real time staging of the performance to emphasize its tangible, transgressive quality. Morphing the design in this way also influences the shifting nature of immediate and subsequent individual interpretation. This is apparent in the design of Carnesky's work, from *Jewess Tattooess* (1999a, b, 2001) to *Ghost Train* (2004, 2008). *Jewess Tattooess* in particular allowed the design to morph from venue to venue, due to Carnesky's artistic (re)writing of the work. It mutated further within each performance due to her physical interaction with the sets she performs within, both the live and pre-recorded film elements and the structural design. In the Copenhagen production Carnesky emerged from a Star of David made entirely from pages of the Torah (Hebrew sacred text), leaving the traces of her body in the set as she issued from it. Later she added to this with bloodied footsteps, imprinting her body upon the already imprinted text. As she scarred her own flesh with the tattooist's needle her body merged with the design of the piece as site, sight and cite of performance. Each Star of David inscribed in her flesh remained throughout the run of the performance, and lasted beyond as the traces of her work in her own flesh.

As these examples and each of the discussions in Part 2 illustrate, the (syn)aesthetic hybrid encourages a visceral impact in design fused with body that has direct consequences on performer and audience, truly leaving traces at each stage of the experience. Traces of the performance are left within the design just as traces of the sensual moment are etched within and upon the individual performer and cited within and upon the corporeal memory of the audience member.

The (syn)aesthetic hybrid pertains to Broadhurst's descriptions of hybridity in liminal and digital performance, 'that simultaneously distances and engages the spectator' (Broadhurst, 1999a: 71). Such hybridity establishes form as 'a merging of the aesthetic with everyday life', 'montage', 'dreamscape', 'collage' and 'imagination', that works to instil 'lasting effects' (77–89). Of particular note here is how, with a visceral (syn) aesthetic hybrid, multimedia, digital and bio technologies can be manipulated to counterpoint and coexist with the live performance, in order to foreground and counterpoint the live experience of the performer/audience relationship.[3] In such situations the presence of Derridean wide jarring metaphors is rendered unusual by the fact that the 'interconnectedness' of the senses is both distanced and brought closer by the technologically manipulated sequences (see Broadhurst, 2007: 30). Curious are particularly eloquent on the sensuality of the technologically mediated, 'visceral-virtual' moment in their conversation in Part 2. Paris states:

> [T]hat word 'virtual' is interesting because on first glance it implies distance but actually it's also about what's palpable, what's 'almost there'...what's absent and present at the same time.

Many of the practitioners highlighted in Part 2, in particular Bodies in Flight, Carnesky, Curious and Graeae manipulate a live recording of sections of the performance as it runs. They juxtapose the live with the pre-recorded, the live with the live-recorded and the live-recorded with the pre-recorded, defamiliarizing the live moment and undermining 'accepted boundaries and definitions' (Broadhurst, 1999a: 24).

Performance practice that executes a complex (syn)aesthetic hybrid demands much of its audience. The audience is expected to experience and interpret a whole stage picture that interweaves live performers with design elements, which may involve various technologies, pre-recorded and live soundscores (including verbal texts), so that divisions

Figure 1.5 Marisa Carnesky in *Carnesky's Ghost Train* (2008). Photo Credit: Marcus Ahmad

between form and content become perceptibly inseparable. All the senses are called into action; that which is visible, audible, olfactory, haptic, tactile and tangible becomes crystallized in the performance format, foregrounding the form of the performance as a semiotic site of transgressive and intertextual communication. This is highly apparent in all of Punchdrunk or Shunts' work or in *Carnesky's Ghost Train* (2004, 2008) where the audience enter an otherworldliness, itself at once disturbing and exhilarating. Each individual audience member has to intuit a holistic sensual experience, communicated primarily through body, space and sound. The signification processes take the individual directly into the physiological, psychological and philosophical world of the characters that they are literally in touch with. In these worlds the audience literally plays in the narratives and comprehends the underlying themes in a fused somatic/semantic way.

A (syn)aesthetic hybrid demands a shifting audience perspective. This ensures that an audience perceives the work in a multidimensional manner. When any performance practice manipulates hybridity to prioritize sensate access, it results in a multidimensional (re)cognition within the audience that draws on the multisensory evaluative capacities of (syn)aesthetic appreciation. This is a feature played with by Graeae, a company made up of performers with diverse physical and sensory abilities, in order to allow total and inclusive access to the work via a variety of means. As the conversation between Jenny Sealey and Glyn Cannon in Part 2 illustrates, Graeae's collaborative hybridized approach is crafted in order to evoke an 'equality' of the experiential for each member of the audience and in doing so takes on a political dimension.

The ambiguous and multidimensional nature of a (syn)aesthetic hybrid ensures a 'complex simultaneity of stage processes leading to the impossibility of producing a single interpretation' (Broadhurst, 1999a: 78). This foregrounds the audience as active participants in the performance experience and the process of meaning making as a result of the interplay of the various layers of the (syn)aesthetic hybrid. All of the practitioners speaking in Part 2 highlight the importance of the individual and intuitive role taken by the audience in each event. For Bodies in Flight, Carnesky, Curious, Punchdrunk and Shunt, this is to the extent that they effectively become co-collaborators in the work. Punchdrunk and Shunt emphasize the fact that the active participation of the audience as voyagers through each performance event is vital to their practice. When an audience is encouraged to experience the various layers of 'meaning' in the work by becoming part of the ludic play at the heart of the form itself, a dynamic curiosity is ignited. As a consequence, this playful inquisitiveness penetrates the content and activates a multifaceted journey of interpretation for each individual audience member both during and following the performance.

The varied visceral languages of a (syn)aesthetic hybrid establish an enjoyable and/or disturbing fusion of the somatic/semantic allowing the double-edged making-sense/*sense*-making process to occur. This ensures the performance event is far more that an intellectual exercise and emphasizes its visceral, experiential quality. The very fact that the individual audience member intuits the whole experience of a work is often responsible for a (syn)aesthetic-sense being affected. This is true of the site-sympathetic work of Punchdrunk and Shunt, the multimedia performances of Bodies in Flight, Carnesky or Curious or the interdisciplinary choreographies of Khan. It is also evident in theatre productions which exploit the potential of a (syn)aesthetic hybrid such as the Royal Court production of Kane's *4.48 Psychosis* (2000b and 2001d), as discussed by McInnes in Part 2. In the latter, the use of film projections of a grainy external world (developed out of lived experience during the rehearsal process), woven into the design of the production leaves its traces on the bodies of the performers in the immediate moment and in the minds of the audience in subsequent recall. Interwoven with the powerful and haunting verbal text and the intensity of the concentrated physical performances, this fusing of technology within the design jars reception and produces a disturbing visceral response by establishing a dreamlike quality. It fuses notions of the real with the imagined, the past with the present, the live with the pre-recorded, and exposes traces of live(d) moments. The overall effect results in equivalent traces of the performance being left within the audience member's corporeal memory of the experience.

As reinforced from the outset, within the (syn)aesthetic hybrid discrete elements may dominate which become primarily responsible for affecting the (syn)aesthetic response in appreciation. In particular and of utmost significance to a visceral cognition is the unusual manipulation of physical and verbal texts as prioritized within (syn)aesthetic work.

The performing body and the (syn)aesthetic style

> [T]he transmission of body signals, opens the way to defining a reality determined by corporeal conventions.
> (Broadhurst, 1999a: 77)

> Whatever can be said of the body can be said of theater.
> (Artaud qtd. in Derrida, 1978: 232)

The predominance of the body is a vital and defining strategy of (syn)aesthetic performance. (Syn)aesthetic work is both signified *and* experienced through the human body. It is this factor that is responsible for the immediacy of the appreciation experience. The body in (syn)aesthetically styled performance is foregrounded as a sensate text which can be read via sense/*sense* impressions and can be manipulated in the (syn)aesthetic style in a variety of ways as illustrated by all of the practice discussed in Part 2.

The human body as a signifier within (syn)aesthetic performance holds great potential for deep penetrating communication. The eloquence of the human body to translate or (re)present internal and external human conflict, simultaneously becoming an integral element in the hybridized form, is a factor that has confronted and overturned traditional 'Westernized' approaches to producing and interpreting performance.[4] An individual body is open to an abundance of readings and enables the play of representation in corporeal discourse. The (syn)aesthetically styled body in performance provides the slippage and fusion between various sensual languages, such as the verbal, haptic and olfactory, which is then experienced as an

(Syn)aesthetics in Practice 63

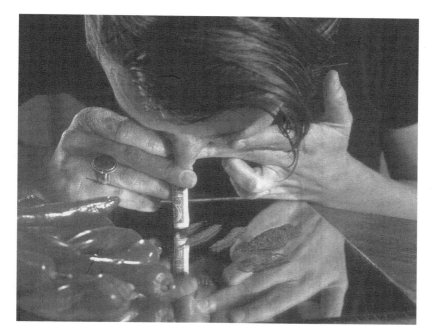

Figure 1.6 Leslie Hill in Curious' *On the Scent*. Image © Curious/Arts Admin. Reproduced with kind permission. Photo credit: Hugo Glendinning

equivalent sensation through the (syn)aesthetically perceiving bodies in the audience. Such a feature is exploited in the work of Curious, particularly *On the Scent* (2003) and *Autobiology* (2008–2009) where the sensual workings of the human body are manipulated within the narrative and thematic experiences.

Artaud's arguments for a 'writing *of* the body' (Derrida, 1978: 191, emphasis original) are made manifest in performance work that utilizes the actual body as the sentient source and conduit of sensate communication. The exploration of marginalized experience within and between actual bodies in performance demonstrates a physical practice of écriture féminine (Cixous, 1993; Irigaray, 1985). Embodied narratives embrace the notion of writing from the margins via a truly corporeal writing of the body. The body in performance automatically displays the 'transformation of each one's relationship to his or her body (and to the other body)' (Cixous, 1993: 83). A play with notions of writing the body occurs in the Bodies in Flight intimate production *To Deliver Us* (2001). In this piece the female performer controls a hand held digital camera, which she has been using to explore her male partner, intermingling their fleshly bodies with the mediated version of the same. In a moment of charged intimacy she plays with and (in)between the camera, interacting with his unmediated body, to write upon his body both with her body, caressing him with her hands, also literally inscribing the narrative of their physical relationship on his body with a pen; 'I woz ere' on his foot 'and here' on his throat and so on (see Figure 1.7).

In (syn)aesthetic performance the polyphonic textuality of the body becomes paramount which highlights the shapeshift nature of the human body as well as its potential to communicate and interpret multiple corporeal consciousnesses. As site, sight and cite of performance, the polyphonic body and its multifaceted capacity for communication is crucial to the (syn)aesthetic style. Furthermore, it is often the case that, rather than supporting or representing 'something spoken', the movement and physical quality of the actual body in performance 'speaks' itself, 'leading to a free association of themes rather than a linear narrative which can provide no answers in manifest or rational (or linguistic) terms' (Broadhurst, 1999a: 65). Thus, 'as direct working material' the human body 'goes beyond the representational role-playing of theatre' (103). Both Khan and Wallace draw attention to the unique powers of communication only attributable to the human body in Part 2.

The body in performance reve(a)ls in its own narratives and simultaneously confronts diverse corporeal polyphonies, or consciousnesses in the expression of marginal experience. This is illustrated in the choreographed work of Carnesky, Khan, Graeae and Punchdrunk and equally in the written texts of Churchill, Kane, and Wallace. Significant to the (syn)aesthetic style is the fact that emphasis is often placed on a very real lived and living body conveying its own history. As Wallace notes in Part 2, where history and social relationships are enacted through the body the audience becomes intensely aware of the human body as a site of performed history and *actual* history; the body as the site where 'power is enacted or struggled over'. In this way, works which foreground the physical body align themselves with the marginal by (re)presenting a 'reality determined by corporeal conventions' (Broadhurst, 1999b: 22). Here it is the actual body that is carnivalized at once subversive, shocking and celebratory.

The polyphonic body, with its potential to represent itself as a site of struggle and conflict, exposes the Dionysian traits of duality, disturbance and playfulness. It can expound the lived experience of an individual (gendered, sexual, historical, political and so on) by ensuring that, 'the entire symbolism of the body is called into play' (Nietzsche,

1967a: 40). A playful performing body makes physical Derrida's notions of iterability marking a 'relation between repetition and alteration' and critiquing 'pure identity' (Broadhurst, 1999a: 50). Furthermore, the (syn)aesthetic body in performance 'tells' an individual's experience of her or his own body which in turn allows the perceiving body in the audience to (re)cognize, in an experiential manner (and via a corporeal memory during and following the event), both the other individual body and her or his own individual body. This is clear in the work of Bodies in Flight, Carnesky and Curious and is especially the case with Khan's work, particularly *Zero Degrees* (2005), *Sacred Monsters* (2006), *bahok* (2008a) and *in-i* (2008b).

The (syn)aesthetic mode of production and appreciation exploits the actual body's potential for deep penetrating communication. The body in performance expounds Kristeva's notion of the human subject as the 'play of signs' (qtd. in Broadhurst, 1999a: 6), fulfilling an individual's ability in performance to transgress and displace linguistic communication by playing with the corporeal potential for signification. A (syn)aesthetically styled body within performance provides a unique access to the 'lived' as an experiential dimension. Following Artaud, the performer's body in a (syn)aesthetic performance is manipulated so that it works 'directly upon the nervous system of the audience' (Ward, 1999: 126). This highlights the significance of the body as the primary interpreter, where the performance is 'aimed at the whole anatomy...unafraid of exploring the limits of our nervous sensibility' whereby performer and audience make a 'substantial journey *through the senses'* (Artaud, 1993: 66, 89, emphasis original). The perception and interpretation of the piece is designed to 'enter the mind through the body' (77).

This is particularly apparent in Carnesky's *Jewess Tattooess* (1999a, 1999b, 2001) or Khan's *Zero Degrees* (2005). In these works the practitioners bodies reveal their historical, cultural, 'personal' and performing identities.

In *Jewess Tattooess*, Carnesky's actual body is presented for the audience to touch, to experience in the flesh, whilst the interwoven narratives of the (syn)aesthetic hybrid engage spectators in the historical, cultural and personal experience on both an abstract and a concrete level. In *Zero Degrees* Khan and Sidi Larbi Cherkaoui recount Khan's story of a challenging journey through an Indian border crossing which is marked by the death of a man on the train. The piece becomes an interrogation of rites of passage, cultural belonging and un-belonging, death as a point of transition, all framed by the mesmerizing music of Nitin Sawney, locking the piece into the sensation of memory, loss and longing whilst Antony Gormley's stark white walls and white sculptures of ghostly bodies simultaneously make the experience cold and alienated and throw into stark relief the sensuality of the dance. This dance (sometimes as duet, sometimes solo pieces with the other performer looking on, witnessing), demonstrates a compelling live(d) vitality. The choreography plays out emotional and physical equations that the body works through in the urge to make *sense*/sense of a complex experience. It is such a manipulation of the live(d) body as site, sight and cite of performance that becomes a point of exhilaration and direct connection and also be the cause of disquiet and disturbance in appreciation of (syn)aesthetic work.

As suggested by the Bodies in flight reference above, interjections from mediated images (pre-recorded and live-recorded) of those live bodies present in the performance plays with form and serves to fuel the immediacy of (syn)aesthetic appreciation.

The technologically (im)mediate body, rather than reducing sensual perception, can serve to extend embodied experience, render it unusual as it 'alter[s] and recreates our experience in the world' (Broadhurst, 2007: 24). This is made manifest, as both presence and absence in Curious' *Vena Amoris* (2000–2007) as discussed in Part 2. With this piece the interaction between individual audience member and individual performer

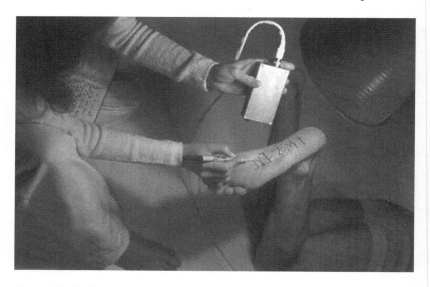

Figure 1.7 Bodies in Flight. *Deliver Us* (1999–2000). Performers, Mark Adams, Polly Frame. Image © Bodies in Flight

became highly charged due to the intimacy of the one-to-one, even though mediated by a mobile phone.

A transgressive body in performance can present those bodies which fail to conform in such a way that the transgressions are *felt* through the perceiving bodies in the audience. Strong examples of such are Graeae's practice which simultaneously destroys and celebrates notions of différance in relation to the diverse types and experiences of the human body; or Churchill's shapeshifting Skriker, as twisted and mutating as her speech (Churchill, 1994a); Wallace's play with the politics of desire via a corporeality that reveals personal, social and historical experience through transgression and transformation, which is played out on and through the body; or the tormented bodies of Kane's work, non conformist bodies which 'transgress' both because they are 'other(ed)' (gay, abused, disfigured) and also because they perform transgressive acts (raping, dismembering, bodies in seizure); Khan's dancing body, constantly working out its (con)fusions of experience and Carnesky's tattooed and tabooed body. Broadhurst, following Michel Foucault, highlights how disturbatory bodies are 'non-"docile bodies"' (1999a: 178) which confront and explode notions of conforming bodies. This rewrites conformity in terms of both the individual body – sensual, emotional, psychological, sexual and so on – as well as in terms of the historical, social, cultural and political body.[5]

Exploring the potential of the body as a site of performance enables a direct experience of Kristeva's semiotic, that which is irrational, unconscious, sensate and transgressive. In this way, the human body explores and presents the dialectic between internal and external experience. This idea is beautifully explored in all of Khan's work where internal and external lived experience is translated through the visual, aural, spatial and corporeal. In (syn)aesthetic performance the body is a means of making the intangible tangible, a mode of communication that enables the saying of the unsayable. It is thus a crucial factor in producing a (syn)aesthetic-sense within the audience. Here the sentient human body as the source of lived experience employs and instils a primordial sensitivity in both production and appreciation. Many of the practitioners in conversation in Part 2 provide evidence of this, in particular, Punchdrunk, Khan, Bodies in Flight and Curious.

In this way, the actual body in (syn)aesthetic performance proves itself to be a chthonic conduit, an experiencing agent for performer and audience alike. The body provides the means by which there is a return to the primordial within fused multisensory cognition, with an emphasis on the haptic and tactile. The (syn)aesthetically styled body ensures that corporeal images, traces and memories, rather than intellectual thought guides immediate and subsequent thinking. Here the primitive sensitivity of the body is called into action by performing and perceiving bodies alike. In engaging a making-sense/*sense*-making strategy of appreciation, the audience appreciates body before and within knowledge, corporeal memory before intellectual analysis, so that rational reflection becomes secondary to visceral experience.

The (syn)aesthetic body in performance articulates those sensations, those experiences of the (syn)aesthetic-sense, that are beyond the powers of verbal communication. It attempts to 'retrieve a chthonic identity by direct corporeal insertion into the creative act' (Broadhurst, 1999b: 22) and asserts a fused body by moving away from the mind/body split, privileging the chthonic *as part of* a noetic reasoning. This is illustrated by Punchdrunk's *Sleep No More* (2003) where the body moving in space ensures the intangible is made tangible. As Doyle explains:

> [T]he performer playing Macbeth was having real difficulties accessing the psychology of Macbeth before he kills Duncan....But as soon as he discovered this room, with spikes all over the walls he was instantly able to find something that opened that up for him...the space offers up more possibilities for the performer to interpret their role beyond the immediate and beyond the studio. It offers both physical and psychological dimensions (Doyle and Machon, 2007; see Figure 1.4).

The visceral experience of the fleshly, dancing body fused with the architectural sensuality of the space allows for such intangible elements as character psychologies to be made tangible. As the Punchdrunk example demonstrates, a sensuous interpretation of abstract ideas is sighted/sited/cited in and on the performer, which is in turn cited in the corporeal and intellectual experience of the audience. Ideas that exist in the text are literally fleshed out, presented and appreciated corporeally, which ensures the making sense/*sense* making process of (syn)aesthetic appreciation is activated. Similarly, in Part 2, Khan also clarifies how certain dancers have an ability to 'become' an idea, to communicate a concept which enables the audience to experience directly philosophical thought as much as abstract emotion or narrative.

The (syn)aesthetic style foregrounds the body in performance as the sentient conduit for communicating and interpreting human experience. Developing Artaud's demands for performance work to communicate with a new language through the writing of the body, (syn)aesthetically styled performance prioritizes the body in such a way that it continuously (re)writes itself as a multifaceted, sensate signifier in order to explore, present and interpret contemporary states, events and concerns. Within (syn)aesthetic performances the presence of the body as visceral conduit and sensual architecture, means that the vitality *of the form alone* becomes a sensate experience enabling a 'special perception' (Shklovsky, 1965: 18) during the performance and engaging a corporeal memory in subsequent recall.

Within the (syn)aesthetic hybrid there is often a special connection of body and *play*text. (Syn)aesthetically styled speech demonstrates an awareness of the human body's capacity for unique, *felt* communication. It concerns itself with a corporeal quality in writerly style and exposes the possibilities for saying the unsayable through a fusion of verbal and physical image. This is illustrated perfectly by the work of Churchill, Kane and Wallace.

Disturbing speech patterns – the visceral-verbal *play*text

> Sometimes there are sentences...which you love so much you want to inscribe them on pebbles, tattoo them on your arms.... Like stones for our minds to ruminate on, to turn over and over in every direction.
>
> (Novarina, 1996: 113)

> [I]t is possible to present the 'unpresentable'... from beyond but also including language.
>
> (Broadhurst, 1999a: 8)

The (syn)aesthetic style reclaims performance writing as a sensate and multilayered form. A crucial feature of the (syn)aesthetic style is this reclaiming of the word, the act of writing and verbal delivery, as an embodied event and a sensual act which take on the visceral qualities of communication. That is, both the ability to stir innermost, inexpressible human emotion and to disturb those viscera which cause aural, visual, olfactory and haptic perception. In this way, verbal and written language itself takes on the double-edged quality of making-sense/ *sense*-making akin to the (syn)aesthetic style.

The visceral-verbal *play*text defines writing that takes on corporeal signification to communicate in a fused experiential and noetic way. In short, language is both a cerebral and a corporeal act, and the cerebral and corporeal potential of verbal texts fuse in (syn)aesthetic work. Here it is the imaginative leaps that the audience are enabled to take with the performance that are fundamental to the immediate experience of the work and any subsequent processes of individual interpretation.

The inheritance of (syn)aesthetic writing can be seen to lie in a fusion of historic writerly styles, in particular; the linguistic lusciousness of Shakespearean and Jacobean writing and the aesthetics of Modernism and feminized practice. Such writing breaks away from narrative and conventional dialogue and (re)writes linguistic conventions in order to (re)present and make sense/*sense* of the social, cultural and political mood of the time. Following this, (syn)aesthetic *play*texts demonstrate a desire to show that internal, semiotized human experience (the subconscious, abject, emotional and so on) is as complex and significant as any external experiences and narratives that are placed upon it. Prevalent in (syn)aesthetic *play*texts is a writerly ambiguity that provides interpretative freedom and disturbatory pleasure in the layers of meaning which explore difficult and complex states, revealing 'polyphonic consciousnesses' (Bakhtin, 1984: 17–18). They also have the potential to allow words to touch the unconscious so an ineffable experience is *felt* in appreciation.

(Syn)aesthetic writing crystallizes and concentrates the intensity of personal, lived experience and themes, revealing the intangible (political ideas, psychological states, taboo concepts) through tangible speech and imagery. (Syn)aesthetic *play*texts connect wider social, historical and cultural issues with the individual and personal in an unusual and evocative way. Their fused visceral and noetic style communicate somatic experience alongside conceptual ideas, which the audience can appreciate through the form itself. This is aptly demonstrated by the sheet sequence from Wallace's *Things of Dry Hours*, as discussed by Wallace and Kwei-Armah in Part 2:

> *In the dark, Cali is asleep on the floor among the largest pile of sheets we have yet seen. These sheets seem to glow in the darkness with a strange, ethereal light. Magically, a sheet rises from its pile and floats in the air above Cali. If possible, more than one sheet rises up, perhaps many. The sheets float around the stage like ghosts. Cali sits up in her 'dream', and marvels at the floating sheets. She reaches to grab one, but it evades her. She tries again.*
>
> **Cali:** Come back here, you.
>
> *Cali follows the sheet/s around the room, mesmerised by it, jumping for it, laughing as it evades her, enjoying the strange game.* (Wallace, 2007a: 51)

Similarly, in Churchill's *Far Away*, the hauntingly disturbing presence of the hat parade:

> Next day. A procession of ragged, beaten, chained prisoners, each wearing a hat, on their way to execution. *The finished hats are even more enormous and preposterous than in the previous scene.* (Churchill, 2000a: 24, emphasis original)

In the original Royal Court production, accompanied by distorted, unsettling cavalcade music, the chilling movement of this hat parade produced a disquieting effect, which draws on traces and echoes of lived historical events. The prison clothes, the bodies chosen (all shapes, sizes and ages – most disturbing the fact that the wearer of Joan's winning hat is a child, clinging on to an adult's hand) instantly evokes the holocausts, genocide, of recent history; Auschwitz, Cambodia, former Yugoslavia, and on. Yet, the presence of the hats serves to push this image further, delving deeper into the dark and dangerous possibilities of human nature. The defamiliarized image engages the (syn)aesthetic-sense and allows understanding of a simultaneously abstract and primordial idea; the notion of the human heart of darkness. It is the manipulation of speech, gesture and image, written into these *play*texts, alongside the other elements within the (syn)aesthetic hybrid that work with and/or against the words that encourages a powerful (syn)aesthetic response to be achieved.

As with the play of multiple discourses available to the actual body in performance *play*writing can juxtapose a variety of linguistic registers, emphasizing the corporeal and interdisciplinary within its very form. (Syn)aesthetic writing can destroy boundaries and cross fertilize itself with other disciplines and discourses, interweaving these within the substance of the text in order to produce a defamiliarized, visceral impact. Churchill's *The Skriker* (1994a), *The Lives of The Great Poisoners* (1998b) or her text for The Siobhan Davies Dance Company's *Plants and Ghosts* (2002c), Cannon's *On Blindness* (2004) or Jones's texts for Bodies in Flight, interweave diverse linguistic registers, shaping them within or around dance, signing, music and design, ensuring these elements exist in the very substance of the *play*text. They demonstrate that *play*writing can be perceived as a physicalized practice in itself, with an indefinable nature and inherent resistance strategies.

The (syn)aesthetic writing style demonstrates how verbal signification morphs its own morphology, becoming, 'something else at any moment' (Irigaray, 1999e: 55). In this way, where traditional approaches to playwriting have been considered to be reductive, enforcing closure in meaning-making processes, with (syn)aesthetic *play*texts (from conception to performance), an opening process is established in terms of appreciation and analytical strategies.

In terms of words touching the unconscious visceral-verbal *play*texts have the power to make tangible the intangible. Here the word itself has the ability to activate the (syn)aesthetic-sense. Broadhurst, following Lyotard, highlights the experience of the unsayable as that 'something which should be put into phrases, cannot be phrased' (1999b: 21). With (syn)aesthetically styled speech, this 'something' *has* been phrased, and phrased in an unusual and immediate linguistic manner in order to foreground that which, formerly, formally denied phrasing. The noetic capabilities of language in (syn)aesthetic *play*texts come about because the audience hears the words first with their bodies, with a primordial sentience, an embodied knowledge. To achieve this the transcendent quality of language itself is manipulated, enabling the verbal act to return to the chthonic forces and possibilities of the imagination. This language is 'the normally *unspoken*' which articulates the ineffable, allowing audiences to 'become party to a secret...share a transgression' (Barker, 1997: 167, emphasis original).

In this defamiliarized state of 'exposed lucidity' (Derrida, 1978: 242), words become integral to the (syn)aesthetic nature of a performance and are responsible for the disturbing nature of the reception. Etched onto the bodies of the audience, the words themselves become corporeal citations in appreciation. This is particularly true of Churchill's *Far Away* where verbal language is manipulated in such a way that the familiar is made unfamiliar, as with Joan's final speech:

> It wasn't so much the birds I was frightened of, it was the weather, the weather here's on the side of the Japanese. There were thunderstorms all through the mountains, I went through towns I hadn't seen before. The rats are bleeding from their mouths and ears, which is good, and so were the girls by the side of the road. It was tiring there because everything's been recruited, there were piles of bodies and if you stopped to find out there was one killed by coffee or one killed by pins, they were killed by heroin, petrol chainsaws, hairspray, bleach foxgloves, the smell of the smoke was where we were burning the grass that wouldn't serve (Churchill, 2000a: 37–8).

In the immediate moment this speech covers the evolutionary theory of the survival of the fittest; cockroaches, wasps, crocodiles, humans alongside the twenty-first century 'apocalypse' of humans against nature. It confronts the danger in separating humanity from nature and the primordial. Most disturbingly this speech exposes the recurrence of culpability, drawing attention to the play's own form within that recurrence (the play begins with Harper ensnaring herself in her own chilling lies to the younger Joan). It makes tangible the fact that humans collude with acts of atrocity against other humans, other species and are culpable in globalization, culpable in pollution, culpable in ethnic cleansing, culpable in lying to children and in believing their own lies. This short speech enables, in the fleeting moment of immediate delivery, and in the subsequent processes of visceral interpretation, a disquieting (re)cognition that, like the resonance of chaos theory also embedded within it, the act of lying to a child can produce a world that destroys itself.

Such visceral play allows the audience to 'experience the play moment by moment' (Barker, 1997: 38) where it is the speech that stimulates an immediate (re)cognition, which evokes the ineffable (syn)aesthetic-sense.

(Syn)aesthetic *play*texts engage words in a manner, 'distinct from their actual meaning and even running counter to that meaning' by way of creating 'an undercurrent of impressions, connections and affinities beneath language' (Artaud, 1993: 27). In this way the (syn)aesthetic visceral-verbal and a writerly manipulation of stage image in directions can 'give words something of the significance they have in dreams' (72). Churchill, Kane and Wallace crystallize imagery and craft theme and narrative to ignite a dreamlike quality in all of their work. With Kane's *4.48 Psychosis* (2000b) the layout is played with, choreographed across the page, and the multilayered language does indeed find the affinities across and beneath the verbal, connecting the visceral and transcendental throughout; 'it is myself I have never met, whose face is pasted on the underside of my mind' (Kane, 2000b: 43). The nightmare world of Kane's *Blasted* (1996) places the audience in-between the real and the imagined, creating unnatural happenings which smack of the violent and troubled world in which we live. Its themes as present today as they were in the early 1990s when it was written:

Soldier: Me?
  (*He smiles.*)
  Our town now.
  (*He stands on the bed and urinates over the pillows.*)
Ian is disgusted. *There is a blinding light, then a huge explosion. Blackout.*
The sound of summer rain (Kane, 1996: 39).

The otherworldly styles of Churchill, Kane and Wallace do not fall into a trap of esoteric fancy but instead pull the audience directly towards lived history and demand a complex response. In this way, (syn)aesthetic writing speaks volumes and opens up meanings in the receiver's mind and body.

The manipulation of linguistic devices such as onomatopoeia, sonority, intonation, intensity and of upturning grammatical rules in a Dionysian fashion, can return speech to its disturbing primal roots, 'its full, physical shock potential' (Artaud, 1993: 35). Here words are employed 'not only for their meaning, but for their forms, their sensual radiation' (83). In this way, (syn)aesthetic performance creates a space where verbal 'images are relished for themselves, and language becomes a sensuality' to counter the 'naturalistic, populist and mechanistic metres of the street' in order to enhance contradiction and disturbance and extol *'the beauty of language'* (Barker, 1997: 88, 114, emphasis original). Such verbal play activates the (syn)aesthetic-sense, engaging the noetic within interpretation, demanding that the imagination is harnessed and actively engaged.

In playing with the fused noetic and chthonic potential of verbal language, (syn)aesthetic writing demonstrates how (syn)aesthetic *play*writing can (re)write speech and expose its shapeshifting, 'infinitely renewable' form (Barthes, 1975: 51). By deconstructing it as an 'understood' semantic tool the verbal becomes a visceral form of communication that releases the somatic essence of words. Here *play*texts explore the border between language and sound, often demonstrating the effects of language at its most damaged and destroyed in order to reve(a)l in its visceral quality. Defamiliarized language, like that presented in the final act of Churchill's *Far Away* (2000), or arranged rhythmically across the page in Kane's *4.48 Psychosis* (2000a), or the play of punctuation and rhythm of Corbin's idiosyncratically remembered communist speeches in Wallace's *Things of Dry Hours* (2007a), demonstrates how verbal language can be (re)played, destroyed and (re)invented in order to produce a more visceral form of verbal communication and thereby find the somatic essence of words and speech.

Words themselves, via their sound and form and their disturbed 'meaning' have the potential to transmit emotive and sensate experience, etching themselves into the perceptive faculties of the holistic body. This ludic disturbance of language can discomfort and unsettle the audience in a sensate and cerebral manner. It causes a (re)cognition of language and allows a reperception of ideas, events, states, experience to achieve a new point of verbal making-sense/*sense*-making.

Spoken language interwoven with other components of the (syn)aesthetic hybrid, becomes a further sensate component within the fused corporeal communication. In Part 2 Khan asserts that *Zero Degrees*:

> is literally for me where text, music and dance, movement, become one...truly a marriage between those elements...There are things that words can say that movement can't say, that you can say through voice that you can't say through the body.

The fused power of image and subtle movement is crafted to the same ends as choreographed dance in (syn)aesthetic *play*texts as illustrated by Wallace's feather sequence in *The Trestle at Pope Lick Creek*:

> *Pace and Dalton are sitting together, a few feet apart. Pace is blowing on a small feather. We hear the sound of her breath in the silence. Pace blows the feather into the air and keeps it above her head, blowing on it, just a little, each time it descends. She lets it land on her upturned face. Dalton watches Pace. Pace sees him watching her. She gives him the feather. He tries to copy her. He does so, but badly. Pace just watches. And laughs. They are enjoying themselves. Then Dalton 'gets it' He knows how to do it. He blows the feather up and keeps it in the air. Pace watches him. Then he lets the feather float slowly down between them. They are both quietly happy. Because they no longer feel alone. Because they are watching each other just being alive* (Wallace, 2002: 333, emphasis original).

In (syn)aesthetic writing, words, via their ludically manipulated sound and form play within 'meaning' and have the potential to transmit primarily emotive and sensate experience. Combined with the performing body and other elements of the (syn)aesthetic hybrid they become images and words that 'lacerate' (Artaud, 1993) themselves into the perceptive faculties of the sentient body. Language in this mode becomes far more than merely aural description. It is able to penetrate deeper as a result of the manipulation of word as somatic sound capsule and semantic sign, painting corporeal images in multisensory ways. Such writing returns human communication to its primitive roots, corroborating the idea that all verbal communication originate via synaesthetic and synkinetic means.

(Syn)aesthetic *play*texts also allow 'words without meaning' to be 'necessary' (Shklovsky qtd. in Eichenbaum, 1965: 109), releasing the performers and audience member into the free-play of their imaginative faculties. By manipulating the vestiges of linguistic meaning, an

76 *(Syn)aesthetics*

Figure 1.8 Zero Degrees (2005). Khan with Sidi Larbi Cherkaoui. Reproduced with kind permission of Akram Khan Dance Company. Photo credit: Tristram Kenton

(Syn)aesthetics in Practice 77

Figure 1.9 *Zero Degrees* (2005). Khan with Gormley sculpture. Reproduced with kind permission of Akram Khan Dance Company. Photo credit: Tristram Kenton

audience can interpret unusual 'meanings' through the visceral reperception of the *sound* of the word. The shifting nature of such linguistic play forces an audience to 'materially recast' (Novarina, 1996: 52) in a corporeal manner both 'lexicon' and 'thought' (Weiss, after Novarina, 1993: 92).

Here, the Dionysian content within verbal language, encourages the 'experience of words' to be 'a measure of their expressiveness' (Luria, 1969: 91). Such embodied and imagistic word perception enables an audience to experience every word, perceiving the details – aurally, visually, physiologically. The semantic 'meaning' of words is therefore reflected as much in the somatic sound and emotional resonance they embody. Via this sentience, words can 'impart the sensation of things as they are perceived and not as they are known' where a 'special perception' (Shklovsky, 1965: 12, 18), shifts 'the signifier' that the words formally encompassed, 'a great distance' (Barthes, 1975: 67). This playful defamiliarization ensures that its making-sense capacity, of a semantic, cerebral kind, is instilled with a *sense*-making capacity of a somatic kind. This is true of the destroyed language and structures of Churchill's *Blue Heart* (1997), *Far Away* (2000), *A Number* (2002) or *Drunk Enough to Say I Love You?* (2006) which get to the noetic and *felt* essence of the situations presented just as it is applicable to the underscored mumblings of distilled and deconstructed text uttered by the performers in Punchdrunk experiences.

The (syn)aesthetic play with the spoken word exposes *feeling* and experience through its rhythms, its sounds, its connotations so that the powers of 'intoxication' serve to 'subvert reason' in interpretation (Barker, 1997: 213). Developing Brik's proposition of rhythm as an integral part of sensate expression discussed in Chapter 2, these playful verbal texts in performance highlight the fact that (syn)aesthetically manipulated language takes on a physical signification producing a visceral experience of the words themselves. Perceived and understood on a sensual level, the aural, physical and intellectual powers of language, as a fusion of sound, emotion and signification, establish a (re)cognized meaning through the somatic/semantic mode of communication.

It is through this play with the physicality of the verbal that body and word find a curious relationship, where the intertwined corporeality of the human body and visceral-verbal language 'speak the corporeal' (Irigaray, 1999a: 43). (Syn)aesthetic writing appeals to the imagination, in an acknowledgement that the sentient body is able to listen to and understand a more imagistic language at a deeper, somatic level. This corporeal aspect of delivered speech feeds into, and

(Syn)aesthetics in Practice 79

derives out of notions of 'writing the body'. It is the corporeality of the word that a (syn)aesthetic writing style explores and expounds, which encourages a (syn)aesthetic perception of verbal language. Such writerly practice highlights a certain antagonism between speech and physicality, whilst simultaneously foregrounding the potential for a symbiotic relationship between the two. Not only evident in the playfully disturbing language of Churchill's work and the brutally tender speech of Wallace or Kane it is also present in the spatial translation of the epic texts at the root of Punchdrunk's work; the physically explored verbal transcripts in Khan's or Giddens' choreography; the prismatic mixture of Yiddish, carnival vernacular, fairy-tales, Hebrew scripture, magic shows and playful, erotic banter in Carnesky's writerly practice; or the signed, danced, described and designed translations of texts in Graeae's work. All draw the audience inward to a chthonic knowledge by stimulating, through word-play, sensations that are read through the body.

(Syn)aesthetic *play*texts can be entirely corporeal in performance and interpretation. They can often stimulate and/or accompany movement based performance as is the case with Cannon's *On Blindness* (2004) or Jones' writing for Bodies in Flight. Alternatively they may expose or arise out of physiological experience such as in Khan's productions or the work of Curious. Paris highlights her commitment to the generation of text from the 'store of memory in the body' in Part 2. The discussion with Curious in Part 2 also acknowledges the profound effects of visceral-verbal texts in generating corporeal memory in the audience. What is clear in (syn)aesthetic *play*texts is that the corporeality in the visceral-verbal text demands to be interpreted through striking physical images, whether moving or still. In this manner, linguistic play uncovers the 'extreme possibilities of language' and actually leads 'beyond the textual, directly into the morass of the body' (Weiss, after Novarina, 1993: 84–7).

Barthes' 'writing aloud' (1975: 66), is of acute importance to the (syn)aesthetic *play*text. Here the experience of the words is carried by the 'grain of the voice', the 'erotic mixture of timbre and language' (37, 66). Such an 'articulation of the body' (66) accentuates the corporeality of the written text, ensuring that it is 'the *speak*' that becomes 'most physical in the theater' (Novarina, 1996: 58, emphasis original). As McInnes and Kwei-Armah both make clear in Part 2, a physical wrestling with a (syn)aesthetic *play*text in the verbal delivery can stimulate a visceral interpretation of the work, which can be responsible for 'a profound reading, ever deeper, ever closer to the core' (Novarina, 1993: 101). Jones talks of his own writerly wrestling with the text alongside the performers' physical work in rehearsal to make it 'palatable';

> literally in terms of the palate of the voice but also palatable for someone to listen to ... it's only by running up against the unpalatable that it becomes interesting and that's to do with tearing apart, or troubling, the difference between sonority and sense.

Such a 'dance of the organs of speech' (Shklovsky qtd. in Eichenbaum, 1965: 109) (re)writes the *play*text via both the act of performing the piece and also in the translation of this within the process of interpretation. In translating the viscerality of this experience to the audience the 'reconciliation of word and body' (Weiss, after Novarina, 1993: 86) truly occurs via the corporeal exchange from writer, to performer to audience. Thus a making-sense/*sense*-making process is put to full effect in the fused somatic/semantic appreciation. This predominance of the senses and corporeality in (syn)aesthetic writing ensures that it is a very real *writing* of the body in concept and form.

(Syn)aesthetics – *re*defining visceral performance

> Theatre is the only place where the mind can be reached through the organs and ... understanding can only be awakened through our senses.
>
> (Artaud, 1974: 182–3)

The complex and sensual manipulation of physical and verbal language in performance work; the embracing of the technical possibilities that video, film, digital and on-line interaction provide; the fact that site, design, light and sound play a major role in a variety of productions and are increasingly technologically advanced, all blend together to ensure the progressively innovative quality of the (syn)aesthetic hybrid. Its intertextual layering of performance languages goes far beyond linguistic analysis and demands the sensate approach of the (syn)aesthetic strategy of appreciation. The (syn)aesthetic style celebrates the physical image as much as the spoken word. It explores the potential of spoken language to affect on a physical level. Its visceral impact is emphasized by the immediacy of the live experience.

To illustrate, interrogate and celebrate the way in which certain contemporary performance practice has employed such a style in recent years, Part 2 surveys a diverse range of contemporary practitioners whose work exhibits (syn)aesthetic tendencies and demands a (syn)aesthetic mode of analysis in appreciation. The following section lets the practitioners speak for themselves and establishes a forum for sharing, between practitioner and audience member, the experiential quality that is at the heart of their performance work. The variety of practices included provides evidence of the diversity of this performance style. Each discussion illuminates the playful approach to performance as much as to interpretation and highlights a tantalizing performative quality present in the (syn)aesthetic exchange in thinking and practice.

Part 2

Introduction: A (Syn)aesthetic Exchange

> [T]hings such as sound, images and the energy of the play, are extremely difficult to describe. To fill this textual void, we have to tell what can be told and try to clarify what drives us.
>
> (Lepage, 1997a: 26)

Part 1 of this book has detailed how (syn)aesthetics surveys a variety of scientific, critical and performance theories to support and present the terms of its own analysis. The following conversations with leading practitioners in the field of performance illustrate key features of (syn)aesthetic practice and foreground the experiential nature of the audience response to such work.

Part 2 begins with a consideration of the (syn)aesthetic hybrid and the sensate quality of space in conversations with Felix Barrett and Maxine Doyle regarding the site-sympathetic events of Punchdrunk and David Rosenburg and Lizzie Clachan on the theatrical experiences created by Shunt. Punchdrunk and Shunt both exploit the multidimensional, multisensual possibilities of the (syn)aesthetic hybrid. These contributors highlight how the audience experience is central to (syn)aesthetic practice. They draw attention to the way in which the audience takes journeys through each theatrical event, which is at once an individual and a communal experience. Both companies discuss the multiperspectival nature of audience experience, which fuses the chthonic and noetic. Central to each conversation is the ludic play that informs the work for both companies, conceptually, formally and actually in the audience's navigation through every experience. Doyle and Barrett also consider Punchdrunk's interrogation of textual sources and how they push the spatial, temporal, conceptual and corporeal dimensions of visceral-verbal texts to unusual and exhilarating ends. Alongside this they draw great attention to the power of the moving body in Punchdrunk's evocative worlds. An interesting point drawn attention to by Rosenberg is the fact that such active engagement with this hybridized work becomes a political event in itself and can be exploited to these ends. Each of these fascinating discussions lays bare the possibilities of space as a constant, tangible prae-sens in performance.

Part 2 then goes on to highlight further the prominence of the body in (syn)aesthetic work, exploring ideas around the body as site/sight/cite and source in discussions with Akram Khan, the internationally renowned choreographer and artistic director of the Akram Khan Dance Company and award winning performance artist Marisa Carnesky. These conversations highlight the prominence of the body within each artist's own idiosyncratic hybrid. These dialogues examine the body as the source of material that underpins each performance and prioritize the stories that the body is compelled to tell; both foreground how the body creates its own language, which communicates, polyphonically, personal, cultural, political and historical narratives. Khan details the need to get to the essence of the individual body in order to plunder this 'museum' of information.

A consideration of the various approaches to writing, directing and performing (syn)aesthetic *play*texts follows beginning with Naomi Wallace in conversation with Kwame Kwei-Armah on their writer/director collaboration on Wallace's *Things of Dry Hours* (Wallace, 2007b). This is followed by a discussion with Linda Bassett on her experience of working on the original Royal Court production of Caryl Churchill's *Far Away* (2000b, 2001). It finishes with Jo McInnes' reflections on her collaboration on the Royal Court premiere and revival of Sarah Kane's *4.48 Psychosis* (2000b, 2001d). Each of these conversations touches on how these *play*texts all allow voices to speak subversively and to penetrate challenging concepts. Bassett articulates the unusual demands that such writing places on the audience in experience and interpretation. Similarly, Kwei-Armah and McInnes both talk of the manner in which these (syn)aesthetic texts allow the performers working with them to transcend, which is then translated to the audience. Wallace is particularly eloquent on the play of the body as a site of struggle and liberation in her *play*texts. As these highlights suggest, these contributions foreground the fused noetic and chthonic quality in each of these works. It is this that touches otherworldliness in both experience and interpretation; an imaginative state which communicates immediately lived personal, political and historical ideas.

Part 2 concludes with three discussions which survey the multisensory and experiential nature of hybridized (syn)aesthetic work with a particular focus on the fusion of body, space, technology and *play*text. Here, Jenny Sealey, artistic director of Graeae and playwright Glyn Cannon discuss their collaboration on *On Blindness* (Sealey, 2004). Their conversation points up the way in which Graeae's collaborative approach is crafted in order to evoke an 'equality' of the experiential. Sealey draws attention to the way in which the diversity of individual sensory ability enables access to ideas and experiences in alternative sensual ways. This is followed by Sara Giddens and Simon Jones, artistic directors of Bodies in Flight reflecting on their working processes during the last 15 years; interrogating the fusions of flesh, text and technologies in their ongoing practice. They highlight the sensual and analytical pleasures that exist in this marriage of flesh and texts, the felt and the thought, for practitioners and audience alike. Jones makes interesting comment on the way in which the physiological body consists of its own system of technologies. The closing conversation with Leslie Harris and Helen Paris of Curious provides exemplary illustration of the multisensory nature of the (syn)aesthetic style and demonstrates the ways in which space, body, text and technology can be manipulated to empower the audience and communicate in a highly visceral manner.

There are a number of points in all of these reflections where the practitioners draw attention to similar characteristics despite a diversity of practice. Each of the discussions suggests the somatic/semantic exchange at every stage of the process. All of the conversations in some way present the play of the body and its multiple discourses, within the interwoven form and content of the works discussed. Doyle and Giddens draw attention to how the dance is *felt* in the performing exchange; haptic reciprocity as much as emotional resonance. Many highlight a certain shapeshifting form, which goes beyond any conventional approach to the rehearsal of a piece of theatre where it may settle and 'perfect' itself over the course of a run. Instead, as these conversations make clear, the transforming nature of the work is something active and tangible that adds to the audience experience. Furthermore, references to the puzzle, game or 'mathematics' of the event at each stage of the process appear throughout these conversations and highlight the play of interpretation central to the hybrid forms explored. The practitioners also draw attention to the mutually beneficial exchange between performance and analysis, each highlighting a fusing of theory and practice within their work.

These discussions articulate how (syn)aesthetics works in practice across diverse approaches to making and presenting performance events. The reflective yet immediate quality of each discussion is intended to draw the reader in to see how (syn)aesthetics can be applied as a sympathetic discourse when analysing all the work under scrutiny. By engaging in a sensual and hybridized approach to thinking/producing/writing/receiving as defined and documented throughout Part 1, the following conversations celebrate the exchange of experience that underpins (syn)aesthetic practice and appreciation.

2.1
Felix Barrett and Maxine Doyle of Punchdrunk: In the Prae-sens of Body and Space – The (Syn)aesthetics of Site-Sympathetic Work

Felix Barrett: A central feature of the work is the empowerment of the audience. It's a fight against audience apathy and the inertia that sets in when you're stagnating in an auditorium. When you're sat in an auditorium, the primary thing that is accessed is your mind and you respond cerebrally. Punchdrunk resists that by allowing the body to become empowered because the audience have to make physical decisions and choices, and in doing that they make some sort of pact with the piece. They're physically involved with the piece and therefore it becomes visceral.

Maxine Doyle: All of the audiences are forced to work instinctively and respond emotionally, rather than intellectually, and often when people take an intellectual approach to the work that's when their frustrations come through. Audience members who've come in with an intellectual approach have to abandon that and follow their instinct. My interest is in the development of a physical language and its relationship to all the other elements. What I've become really aware of is that when you have the audience in close proximity you don't need to compromise on the level of physicality or intensity of your performers; if anything the performers can really play that up. There's something really exciting when you see an audience moving sometimes as a flock to experience a moment, heads ducking to dodge and avoid performers, they become part of the choreographic landscape. My big frustrations with dance within an auditorium as an observer is that with certain work it can feel really distant and theatre spaces can feel really cold so you can't *feel*

Figure 2.1 Punchdrunk's *Faust* (2006). L-R: Fernanda Prata and Geir Hytten. Image © Punchdrunk. Photo credit: Stephen Dobbie

the dancing and in Punchdrunk work you *feel* the dancing, you feel the breath and you have a visceral response to it.

Josephine Machon: You've referred to that instinctive audience response as being Darwinian.

MD: Which is one of the work's strong points but can also be one of its weaknesses. There can be a sense of the audience really relinquishing inhibitions and losing themselves in the world and in the play. If you're less confident, we don't know what audience members bring with them to our world, their history or circumstances, but within that experience you can be overwhelmed, put off by it.

FB: I've always felt that the mask was the one thing that removed that sense of trepidation, whatever baggage you're bringing in, it's neutralized by the mask. So you can be a timid person, but be crazy in the show world.

MD: Yes but I think the mask doesn't necessarily stop you being timid physically; there's a difference between someone who will charge down a dark corridor and someone who will teeter one foot in front of the other because the physicality of the space becomes really intimidating to them.

FB: In terms of being Darwinian there are different routes you can evolve towards. We always establish an entrance point to the world we create, which is like entering a decompression chamber, to acclimatize to the world before being set free in it. Then you can acclimatize to the rhythm of that world. People are really proud when they then make a conscious effort to leave the crowd, or avoid them, and that's some minor epiphany for them. In doing that they've really owned their journey and the moments of performance they find are all the more powerful for that, that was an awakening for them to do that. We're hoping that at some point all of the audience do have an awakening and in doing that they come to realize how they want to address and approach it.

MD: That's the brilliance of the concept actually, and I can say that because it's Felix' concept, it's a constant challenge for the audience, it's never going to be an easy ride. But if they really invest in it and work for it then it can be a really liberating, really exciting experience.

FB: We deliberately don't spoon-feed them; we make it difficult for them to crack it. If there were a set of rules we could tell them how to do it, then they wouldn't have that moment where they say, yes, I've unlocked the puzzle.

MD: In saying that, we have put lots more signifiers in place to lead the audience, to give them more clues, not to follow a narrative but in a really simple way to indicate where action is going to happen or where there's going to be some big shift. Whether that's through lighting changes or music changes, using all the conventions you find in theatre but using them to give the audience clues so that there is some information to help them crack the puzzle. It's all visceral and emotional; we don't really give them any intellectual clues.

FB: That's because we want them to make the intellectual connections. Initially it should be a sprawling wilderness that you can't fathom, you don't know how big it is, you don't know where performers are, you don't know what's happening, you don't know where you are, even if you know the story, that prior information, you don't know where you are in the arc of it. So knowing the rules, working the rules – if you want narrative you follow a performer and you realize the ratio of audience to performer dictates how 'relevant' the narrative is – you can choose which sort of audience member you want to be.

JM: The process of becoming aware of how you're engaging with it becomes multidimensional. You realize that if you unlock one way of working you discover major narratives; another way explores the minor narratives, another, the playful world around those narratives; alternatively you can follow conceptual or thematic ideas by following the space, the design the visuals, the sound. So you can play between the different ways of experiencing the work.

MD: In terms of it being Darwinian, I think that's where confidence comes in. If you're really confident, and I don't mean in terms of personality I mean in this kind of world, in these conventions, then you can really play the game. You don't have to be an experienced theatregoer to play the game, it might be that you're a crazy clubber, or really into gigs so this dark, strange, gothic world is familiar to you and you approach it in a certain way. It might be that you're a really avid filmgoer and the filmic is the thing that stimulates you or you're a visual artist so you're looking at the space in a particular way. It is multidimensional in terms of what people can tune into.

JM: An aspect that is central to your work is space. Felix, you've referred to it as site-sympathetic as much as site-specific. Would you expand on that particularly in relation to the way the sensual quality of the space has influenced the shaping of each different piece.

FB: The space is all-important and the way we build the work is about our instinctual response to it. So it's true, you don't have time to think about it, you let your body dictate to you what the show's going to be. The most crucial part of the process in terms of building the show is the first time the team walk around it, because it's then, when you're wary of a certain corridor and you're tempted by a certain staircase, you know that's your body talking rather than your mind. That's immediate, that's what your senses pick up on. The challenge with the space is to log those feelings and then fix them and accentuate them so that we can guarantee for any audience member that they'll feel that same impact. All we're doing is just harnessing the power of the space, making the building work to its potential. When we plan shows, I deliberately step back from doing too much research on it until the space is fixed because it would completely change the sense, the feeling, the narrative implications of my response.

MD: I go on a completely different journey to Felix in that I don't immediately work so instinctively with the space; in fact I always spend the first three weeks walking around the building being completely lost. I tend to focus a lot on the concept of the content. I do a lot of research and I create a lot of things already for myself. What the building gives me is framings, so I start to see things in relation to framings. When I'm in the studio with the performers we're creating the language in a neutral space but I know where that language is going to be located within that world. During the rehearsal process quite often I'll say, 'don't worry about that, the space will solve that problem', it's almost a cliché but it really does. Felix and I do work really differently until we get on site and that's when everything is fused. Some spaces do inspire you to move. In fact when I'm walking around the space that's what I'm doing, I'm probably swinging off banisters and peering around corners and looking at where we can climb up things and hang off things.

FB: I'm always aware of those places that are good visual points. I always put myself as an audience member walking through. We approach each project as 'one building builds the show'; it is 'specific' but it's sympathetic in the way it's devised. The thought of transferring *Faust* to a different space, different country, different cultural references is daunting. That was really interesting in terms of knowing what the finished work is but prowling a space in order to –

MD: trying to have some spontaneity –

FB: it's really difficult. So much of it will change.

MD: You can't replicate. I think we can transfer and develop but you can never replicate. Our first experience of working together was on *Sleep No More*, we didn't know each other's work at all. We were paired together by Colin Marsh our producer who knew both our work and thought we might be useful to each other. That was a case of creating something in the studio and then moving it to the space and working really easily and organically without too much labour.

JM: As much as you have different approaches you've both talked about how space works out the conundrums, allows you to find the imaginative way in to the concepts and narratives underpinning in the work.

MD: *Faust* for me is one of the most obvious examples, you walk around this empty warehouse that had this darkness about it, which was fantastic and this epic scale but there was a lot of building to do to create this world that we'd envisaged. I remember walking through the space and not really having a sense of what would be where, until we got down to the basement, which was the one space that was suited to hell. It matched beautifully the ideas in some of the final scenes with Gretchen when she's imprisoned –

FB: we both had that same instinctive response –

MD: we didn't change that space, we talked about curtaining it, we talked about flooding it, dramatic ideas, but ultimately it didn't need anything and all we did was put in some simple rigging, which was invisible, and two black straps to do some very simple aerial work. The atmosphere, the space was a given, it felt like it was a gift.

JM: Alongside the fact that you literally descended to reach the culmination of the plot.

MD: You should talk about your 'crescendo' Felix, something we discovered with *Faust*, the crescendo of the space, the crescendo of the show.

FB: We always think about what's that final moment, we always want to tease, tease, tease and pull the rug from under you so the next door that you open is not at all what you were expecting. We realized that in the same way that the audience descend down to hell, which is the logical final point to the building so we wanted the show to filter into that so that the show echoed the building, the show *was* the building. It's also clear with *Masque of the Red Death*. We knew the space as The Battersea Arts Centre which made it difficult to remove the sense of 'BAC' but in terms of exploring it as an empty building, 65 rooms, you've got this giant ballroom tucked out back that a large part of the BAC audience didn't even know about. We knew that this giant space which was, visually and architecturally, the most interesting, just to savour it; it is the crescendo of the building. We could have used it completely differently, created glorious solos. It can be amazing to have a giant space and just one spot with one performer. However, in listening to the building, allowing the building to inform our decisions, it was kind of obvious from the beginning that the audience had to have that revealed; they almost think it's over, it's like a series of doors that get bigger and bigger and if we'd have played our hand too early by opening up the ballroom the building would have felt too open. If you save it until the final point, funnelling the audience together, allowing the audience movement and manipulation through that crescendo of the building so that the audience motion through the space and narrative come to the same point within the building.

MD: What's interesting though, if we compare the crescendo of *Faust* with the crescendo of *Red Death*, is that in Faust the crescendo happens organically without us having to manipulate the audience at all and the space is much more metaphorical. It's a dramatic crescendo and a metaphorical crescendo and it's a diminuendo in terms of bodies, in terms of what you're left with because you're just left with this single image. *Faust* was much more subtle with all of those devices. Whereas *Red Death* is more representational, there are no metaphors in the space in that final sequence, the space is what it is.

JM: I'd like you to talk more about these fusions of form, in relation to the crescendo, the many 'texts' in the work, whether that be prose or play, or a cult film or film score and of course there's the space and the design, the light, sound, the bodies, all of which have a sensuality at their core.

MD: It's fusing things within a really specific aesthetic, a set of boundaries almost. Felix sets them and I push them, sometimes. Often my approach, if I'm not working with Felix, is much more eclectic and I work a little more with juxtaposition and things clashing.

FB: And Punchdrunk deliberately follow one path and really push it.

MD: There's a sense of the homogenous with Punchdrunk shows. It might be interdisciplinary with lots of different elements but there's a sense of it all working together.

FB: It's because it's all about being able to *feel* it. Not just seeing it or hearing it, it's about what the audience feel, because it is about the visceral, so it may be that the most effective way to do it at any given point is a lighting cue, or a performer grabbing you if it needs to be tactile, or if it's the smell of rose petals on the floor, it's about the most efficient way to cut straight to an audience member so there's no hierarchy of discipline. Each one needs to be as strong as the other; otherwise it becomes the weak link.

MD: In *Masque of the Red Death*, there was a debate as to what shadow puppet show we'd create to go in the shadow puppet theatre. I was working with the puppeteer and the performer to create this show and I really wanted to use something different to a Poe story, the Audrey Niffenegger story, *The Adventuress*, and Felix wanted to use [*The Fall of the House of*] *Usher* and we fought about which one would work. We went for *Usher* and it did work better. I was interested in having other associations, not being completely pure with Poe.

FB: But it was because it was very difficult to get the *Usher* story in the main show, so it was just another means of layering. There's always a danger that each show could come across as too busy because there's so many different minds at work within it, so each discipline needs to be working as equal, rising pedestals to support the structure as a whole. The way I saw it with Poe was that there were so many different stories occurring, why bring someone else in when we're lost in his canon.

MD: Although we did end up interpreting some of the Niffenegger text for Madeline because the Poe text was so turgid spoken,

FB: Also the biographical knowledge of Poe himself alongside his work, his fragility and the fact that he's putting all his foibles out there.

JM: Playing with text, paring the verbal element of it right down is key to your aesthetic, from *Sleep No More* onwards.

MD: Poe is meant to be read, not read out loud but his writing engages your imagination so that you create your own relationship with the words. Whereas with Shakespeare his work was written to be performed and the characters are so rich and there's so much vulnerability there that's what I respond to, I respond to the poetry, they're poetic texts. I read them and cut through them to try and get to the essence of what the text says to me.

FB: It's about feeling again, you *feel* the text and you translate that to movement. Both of us always need something to respond to, to bounce off. It's having the backbone set from day one. Because we're so flighty with our ideas we always need that solid backbone to come back to. It means you really can fly away –

MD: because you're rooting it in something.

FB: The reason why we've used Shakespeare so much is because those descriptions, there's so much in there, so many moments, installations are described within the text it's just a matter of unpicking. It's almost like a logic puzzle in itself. You dissect it and then think, this little couplet here will look beautiful in a cupboard, this passage here is obviously core narrative, this bit here is describing atmosphere so that's in the lighting.

MD: It's quite magpie-esque – there's so much there. Looking at what the written text gives us, you can be quite brutal pushing though it to hone in on great scenes, great situations.

FB: If I look back at my complete Poe edition, I went through it once for narrative, then for lighting, then space, then sound, so it became a manual by the end.

MD: One thing that I'm always looking for and that connects all the texts that we've used is something about the human condition. All the work I do, all the work I'm drawn to, there has to be an element of light and shade in terms of the human condition. All of these texts say something about the way that we live and how fragile we are as individuals. So the intellectual experience is absolutely there if you want to seek that out. Our approach is really instinctive, really emotional but our structuring process is really intellectual. It's mathematical actually; it's really dense in its organization. It's great for us when people say that they feel like they were the only person having that experience and it felt like the first and only time that event happened in that way, when in actual fact it's happened hundreds of times over the course of a run.

It's quite a challenge to make work feel like it's spontaneous and happening for the first time and it's because the structure and the organization is really rigorous.

JM: Would you both talk more about the eloquence of the moving body in space, specifically in relation to discovering, what Max has referred to as, 'the unseen words' in the text.

FB: In terms of the difference between the dancer/actor approach, actor's tend to work with their head and dancers just 'do'; their response is immediate and from their bodies. Whereas with actors, there's always a beat before they respond. As a result it doesn't tune at the same level. It takes ten times the amount of work to coax an actor into that world. Dancers, physical performers, build from the floor up; they're living, breathing, one and the same with the space from day one.

MD: I think it's that sense that dancers are really intelligent and their intelligence is located in a different place, it's visceral and embodied, which isn't to say that it doesn't come back to words and discussion. What is magical with dancers is that they're absolutely prepared to just dive into the darkness and not be frightened of drowning. We do also have performers in the company who come from a really strong acting training but who love the approach of play. We're really trying to create a language and my training and experience has always been about trying to create a language of the body that communicates something. Sometimes in our work that needs to be really specific or it can be remote and sensual to just add a layer or a texture.

FB: Before I met Max, Punchdrunk were condensing text, stripping back Shakespeare, cutting up the text and using vignettes. What happened was the audience would be there, tuned into the pulse and rhythm of the building and the work then as soon as the text kicked in, was delivered, they would go back to autopilot, back into being an audience member, and physically their response would change; they'd become more self-aware, they'd shuffle they'd move out the way, they'd be aware of other audience members. It broke the spell. Then with *The Tempest*, we had one dancer who played one of the goddesses, it was very gentle what she did but in terms of what audience members could read into it and with their working knowledge of the text you could see a million things inside this one performance. What Max does so amazingly is create that ambiguity. The movement alludes to something then it shifts and it melts so you can go anywhere with it but also you can pull out direct narrative. A whole passage of Shakespeare can be condensed to one word through that movement.

MD: It is about finding essences and you do have to work with a very particular kind of dancer. The dancer's we work with are actors; they have to have a real ability to tell stories with their bodies and to make movement meaningful. We do work with performers with an acting training and what's interesting is the crossover of performers. The actors bring an intellect and response that pushes the dancers' response in a different way so you get this lovely melange of choices.

JM: The experience of the ineffable is something which is so present in your work and it's something to do with the body and the space and the audience coming together and unlocking something inarticulable in one moment. The intangible suddenly becomes tangible, felt, present on a haptic level.

FB: We've talked about the unifying sense of the work so that there's an aesthetic where all the different disciplines work together. In terms of what flavours the performance in that unusual way, it goes back to the space. The cast are not allowed to see the building until we're quite heavily into rehearsals, so they've already got a working knowledge of their character and the shape of it, but the actual space itself they experience for the first time as an audience member would. They play for three hours, hide-and-seek and performing a series of tasks. They explore it and leave their traces as they go. In doing that, they all get a sense of the bubble in which we're working and that helps to shape the whole piece. There's something there that places our take on the story; something that you can't really articulate but because they've all participated in this game then they all know the pressure of the building and the decompression, the air pressure, the climate. You could never with words say this is the way we're going to shape it, the building defines it. So part of it is a space thing, the ineffable quality, because we never explain it, they just receive it.

MD: The space feeds their presence. Sometimes people talk about our style of acting as being melodramatic –

JM: it's more an intensity in performance than a melodramatic style.

MD: Yes and it needs to be intense, it's essential that their presence is felt and because they need to be able to pull focus when they really want it and sometimes that's really tough when you have large audiences and lots of different distractions. Then sometimes they need to be able to almost disappear into the shadows and not be seen, or they need to pull the intensity right down so that they can have this really intimate moment with just one audience member, which might result in a kiss or the sharing of a secret. And that whole notion of 'presence' is indefinable.

JM: The derivation of presence is prae-sens, 'that which stands before the senses' so that fits entirely with this idea of the live, holistic perception of a moment.

FB: That's great.

MD: That's completely it, and that's a quality that we look for in our performers, an ability to radiate.

FB: It's certainly something that's not spoken. In our audition process you can usually know within the first five minutes who is right for us. In the same way that's how we're trying to engage our audience, so that they absorb it, so that they don't have to be told but work on a primal level.

MD: I remember talking to one dancer who came as an audience member, it was a lovely compliment to the company really, she talked of the generosity of the work in all its dimensions; from the performers in terms of how much they give and how transparent they are, to the generosity of the space and the freedom that the audience is given. We do trust our audience, we trust our performers.

FB: The tactility of each piece, being able to explore the rooms, pick things up, smell things, we have to be generous with that. And the balance of all the elements has to be right, the lighting levels and so on, which is just about us feeling whether it works or not otherwise it's like a museum, not an experiential world.

This discussion follows previous conversations with Doyle and Barrett, which can be found in *Body, Space & Technology*: http://people.brunel.ac.uk/bst/vol0701/home.html.

Details of the performances referred to can be found at: http://www.punchdrunk.org.uk.

2.2
Lizzie Clachan and David Rosenberg of Shunt Theatre Collective: A Door into Another World – the Audience and Hybridity

Lizzie Clachan: Our collaboration is between ten very different people and that diversity, it's the difference that makes the often jarring and bizarre work. You would find it very difficult to do as one person, or even two or three. Sometimes we have little digs and laugh at what each other is interested in. Maybe someone will have an obsession which everyone will roll their eyes at every now and again but it is that little obsession that one person has which keeps things different and odd. We all have our themes because we all come from very different backgrounds, different disciplines so everyone brings with them an 'I need to shoe-horn this in somehow', which is good. People can argue their passions into our shows and that's what makes it interesting.

David Rosenberg: Even though people have different disciplines with which they would place themselves the work always starts as a group of ten artists without any roles, some obsessions, but not roles. As the process continues *then* roles become more apparent. Some people will be performing, designing or whatever, however the work is always created initially from those ten artists.

Josephine Machon: In terms of the 'jarring' you refer to, if we think of your work as hybrid, would you say something both exciting and disturbing grows from the fusion of approaches and perspectives.

LC: Yes because sometimes you can get too neat when you think well this will go well with this and that will work well with that. Often we end up in a situation where we have these disparate notions and actions and ideas which will come together or are around to work with, and that creates something exciting.

DR: I don't think in any of our projects we ever really know how a performance is going to end. There were a couple where we were following more of a narrative line, we had more of an idea of how the starting material ended, but we've never really known how we were going to end it. We would often be left with a collection of the disparate ideas involved, some about form some about content, some about a particular image, and then part of the process becomes a puzzle of how these things fit together and what can come out of that. That is one of the very exciting things about making Shunt work because there never is a prescribed route. There's never a product we're hoping to achieve, there's simply the working through of this stuff, then seeing something with an audience already mixed up in it and then thinking how that can be developed.

LC: For me as a jobbing designer the process outside of Shunt is often very different; it's a non-collaborative, non-'artistic' approach. I work a lot with new writing so it's very much about me trying to find a visual interpretation for what the writer has written.

DR: For me, there are similarities in that, in doing projects on my own whether they're small pieces or full length shows, I still usually piece together a performance from some disparate ideas. However, it's not such an elaborate puzzle as those ideas have all come from one person. In addition to that I don't have to justify the way that those things are put together because they're put together to satisfy me, which is very different to the Shunt projects. They're about how we piece a performance together based on lots of different perspectives because the ten of us are observing the work as it's developing.

LC: Often we can't see what we have until it's in performance. One of the many things you do David is you are the person who eventually has an overview of what we have and what should be chucked out and what might actually be interesting. Usually there's a huge amount of shite which has been very useful because you need to have that to get from here to over there, to make that journey, but in a finished piece, as finished as we ever get, certain material might be superfluous. That element of it, that role, is vital. We do edit over a period of time so there is a situation where people who see our show at the very beginning will find it quite different later on. We've periodically talked about whether [Shunt] have an artistic policy and we could never agree on anything other than 'to explore the live event'. So that gives a good indication of how much the senses, the liveness, is key.

DR: It's to explore that which is unique to the situation of having an audience together with a performance in the same space – and that can be applied in many different ways – but that has to be at the heart of it. The audience being a part of the performance, or a character in the performance, having a role in the performance, is always a very important part of how the work is conceived. I think an important thing about the Shunt aesthetic is that there's often a functional approach taken to the design because there are experiences that we want to give the audience, or ways that we want to place the audience in relation to the action or ways that we want the audience to see the action or see each other. So the devising is always about how the audience sees and moves through and interacts with the space and performance.

LC: It's important to point out that the aesthetic is much more of a collaborative approach when we get down to divisions of labour and to the function and space carving, that may be something that I get very involved in but there is an aesthetic on top of that which comes very much from the ten artists.

JM: Would you talk about the space and the audience and the fusion of theses elements with bodies and texts whether written, verbal or technological; the key areas of sound and space and darkness and scent. What exists for you within those exchanges?

DR: In order to have a group journey the sound has always been a very important element of how to bring the audience through a piece together because they'll be looking at different things or they'll be in different spaces.

LC: We've always had our own space to a degree and all those elements are part of it, there is always going to be video for example, we have projectors and we work with video artists and all those elements are here for us. The video technology is as much a character, a part of the devising process as a performer or as a designer is. These are all elements that exist and express ideas in multiple ways; through sound or technology or a live performer, we'll just try out ideas via various means.

DR: With each different technology there are different ways it works. For example, sound is something that will be continually developed and then changed quite a lot over the course of rehearsal and throughout the development period of the performance, because it can, you can change it quite easily without knock on effects for other areas whereas with video that's always a problem with how the use of video works with this kind of process because it has much more of an impact on the overall aesthetic or design of the piece. It's also not so easily changed in order to keep up with the development of the show. So with each element there'll be a different progression of that part within the piece.

LC: And that is a reason why sound and light are so key; you can change the light and you can change the sound –

JM: They're more organic, they're more like the dancing bodies in terms of the way they mutate and morph with the piece.

DR: Quite often the actual action is usually something that's developed last of all. This has definitely been our process, to open the doors get an audience in and then slowly begin to unpick the mess that we first showed them, but there are very positive things about that process.

LC: I think also that is good or bad depending on our audience. For example, in the old days we'd have our local Hackney audience who would be supportive, they'd understand it and they'd come back –

DR: and that would be exciting for then, they'd want to see it at its rawest –

LC: They're really healthy that audience, that's something we nurture. The flipside of that is then opening *Tropicana* to an almost exclusively National Theatre mailing list audience who probably were just totally shocked by the work.

JM: That relationship perhaps feeds into the element that's present in all of your work which is encouraging in your audience an appetite for curiosity and mystery; developing the work as it progresses means that you're also curious to see what it might become. I'm interested to find about the commitment to the visceral, the provocative and the disturbing in your work, the assault on the senses and the imagination. David, you're clearly interested in pursuing the active role taken by the audience and their 'whole body' experience of the work, how much of your background in medicine feeds into that?

DR: I think for me the most interesting thing in terms of medicine or science in general is in terms of an evidence base for opinion; I'm interested in trying to find out how something works or really 'is', trying to do that by the most appropriate methods. How that relates to Shunt's work is that we don't deal in judgements or dogma or rhetoric.

JM: I assumed there was an interest in human perception and experience specifically in relation to sensation, particularly with your background in anaesthesiology.

LC: You do bring stuff from your medical world and sometimes you bring it in that curious way and sometimes you bring it in terms of the way you make things happen which comes from that work, for example in order to have wine dripping into a glass in *Dance Bear Dance* throughout, which was a brilliant device, that came from David and his blood bags at the hospital.

DR: So access to kit is one area that I bring.

JM: And surely a curiosity about visceral experience?

DR: I don't know whether it is. I'm very interested in medicine, I'm very interested in the relationship a doctor has with patients, I'm interested in the experience of being an anaesthetist and the various responsibilities that are involved in that and of course all of these things make up part of who I am and it would be difficult to separate those things from my work now. Yes, clearly they influence each other but I'm not sure that there is a particular appreciation of the visceral that I try and explore in an artistic way, although there will always be various references – and equipment. The sensual experience is absolutely key for us; I'm just questioning whether there is a connection between that and my background. This desire for a multi-sensory, visceral experience is something that we all share and that comes much more from our overall goals to make work that can only exist in this medium.

LC: I think it's something that we've developed as a company over time –

DR: and it's something that I'm more and more interested in. Our first big show that we collaborated on was *The Ballad of Bobby François;* before that was *Twist*, which was the show that kind of formed the company, but with *The Ballad of Bobby François* this was an auditory experience, an experience of a plane crash in complete blackout, deafeningly loud, it was so loud it was something that your whole body felt. That, together with an aeroplane fuselage being stripped in the blackness around you and all of the things that you felt from that, was a hugely powerful multi-sensory experience and one that was quite terrifying.

LC: *The Ballad of Bobby François* is very important for that [the multi-sensory approach] because we had the idea of doing the plane crash but I don't think any of us realized what we'd created. It was very exciting and everyone who experienced it was very excited by it. It opened up the possibilities. There was an element of stumbling upon something, seeing what we had and then really going for it.

DR: Another element within that crash which was a feeling from within it of being in a larger space that you couldn't yet see because people were confined to that fuselage; that was probably a temperature thing, it was probably warmer in there, and it was pitch black and with the opening out of that, created this other *feeling* –

LC: a really different world and a really different type of performance style followed it, a very strong juxtaposition.

DR: *The Tennis Show* our next show, a show where the audience was split into men and women, that was using smells amongst other things to create a feeling, a state of mind in the audience; certainly from the men's perspective, coming into this Ralgex, Deep Heat smelling kind of homoerotic environment.

JM: It seems that you intentionally frame perspectives within the audience; you shape, to a degree, the different experiences to be had, as with *Dance Bear Dance* where –

DR: there was a revelation of a completely different audience opposite you at completion on the other side –

JM: you're intentionally playing with how individual members of the audience are going to experience the work. That illustrates a shaping of experience and interpretation and the ludic, subversive play that's at the heart of the productions.

LC: We always have a discussion about how much we want to tell the audience what to do. The range might be from, do we get the audience to put on hats through to do we tell them whether they can shout in the middle of the show or not, whether we control them. That discussion is constantly being explored.

DR: With each new project that's a new question to be addressed or decision to be made about how we want to do that. We've always been interested in how an audience, even though they might see different perspectives of something, how an audience could feel, could still experience the whole show together. It's not an individual journey, they might have an individual perspective and of course they will necessarily have an individual interpretation, but all of the audience are on the same journey. That's very important to us but also it's what makes our shows so complicated to put together. It is a very complicated thing to be able to find ways of moving an audience without them feeling like they're being herded; to try and get them to see the things you want them to see in the order that we want them to see them, whilst still allowing the audience to have a role, some sense of freedom. It's very difficult. That's what's interesting about Punchdrunk's solution, it's a very open one, they hand over responsibility to the audience, which works very well in terms of being able to navigate themselves through the installation. That's one solution to working in these more complicated spaces that are made up of a number of parts. We tend to *make* it complicated for ourselves. In terms of *Amato Saltone* it was essentially six groups of audience divided into different spaces, seeing things from different perspectives while we're still attempting to keep the same journey going for all of them.

JM: With that piece in particular, it felt like you were playing on that idea of *playing* with interpretation; one part of the audience would be led around the space and hearing a situation whilst another part are actually viewing it.

LC: I think that what we're playing with is at the edge of where theatre and visual arts come together. The Punchdrunk solution is almost like being in a gallery in some respects. I think we're deliberately interested in the rhythm of a performance and the choreography of the audience through that. I think that's a really important part of our work. In suggesting a way in which an audience moves through, there is a decision there. In leaving an audience to their own devices most will just do what they want and some may get a bit lost as well and never see stuff you want them to and enter into their own time and their own rhythm and we're interested in *creating* a rhythm as part of the act of theatre.

DR: It's a very interesting question regarding the handing over of responsibility to an audience. If there's too much responsibility given – and of course there's no empirical value or amount of responsibility one can give to the audience for any one performance – then an audience can choose not to encounter some of the difficulties of that performance; an audience can choose not to see the thing that is going to upset them or confuse them, or surprise them, or revolt them. It's a very difficult balance because also you don't want the audience to feel manipulated. You don't want to pretend to give them a choice and then actually there isn't one or we're forcing them to feel complicit, we're forcing them to make decisions that they wouldn't necessarily make.

LC: With *Amato Saltone* there was this idea that the audience would come into some kind of key party but we know full well the audience know they're in the theatre, know there's not going to be some kind of orgy. We don't want the audience to think that we think they might think that's going to happen. We know that they know so then we can all play that together. So we often deal with situations where it's audience who are like a congregation –

JM: shifting between individual and communal experience.

DR: Shunt Lounge is a complete exercise in free will. There we see the mayhem of society. With the Shunt Lounge that's where the audience come in and explore the space in the most free way possible.

LC: As an example that is between the two areas, *Dance Bear Dance*, around the table, there was a whole journey of how to deal with audience members who came in drunk and just wanted to shout loudly around the table which just wasn't interesting for anyone.

DR: *Dance Bear Dance* had two main incarnations. The first one, around the conference table, there was much more freedom and points of audience interaction. How we imagined that was that there would be a more free discussion around the table with the audience. The second version we tightened that up much more so we had much more control over that discussion but in fact, as an audience member, because we were able to control the flow of the performance much better, I think it probably actually seemed that they were freer. So there's a very interesting thing about an audience's *perception* of their activity, or their role, in performance or their responsibility in performance and *actually* how that relates to how we shaped the performance and what we know were giving them in terms of the amount of freedom or opportunities for interaction.

JM: How did that play itself out in *Contains Violence* in relation to passive/active voyeurism?

DR: There the opportunities available were very structured. The audience had their seat, they were very alone in terms of being removed from other audience members and their decisions were basically to look for the action where it related to the sound, to look for the action that was going on that didn't have sound related to it or just look at the pissed people in The William Morris Pub over the square; which on some nights was probably the most interesting view and sometimes there were more interesting things happening in other windows. Aside from the overall feeling of power, being perched up on the Lyric terrace with a pair of binoculars, to be able to look out at London, and of course it may be more interesting to look at people who wouldn't imagine they were being watched, that's the excitement. Although within the context of that performance, doing that wouldn't have added to the experience for very long.

JM: That's the interesting thing about this type of performance work as an audience member, as much as you might be utterly immersed in the experience, I think there's an degree to which you are always aware of *the process of becoming aware of* where you are looking and what you are experiencing, the decision you are taking or not taking within that.

DR: That's always the case with the shows, as people are coming in you feel this palpable excitement amongst the audience in just looking around them and deciding on who the performers are. That used to happen in *Tropicana* a lot, you could see people deciding that other members of the audience were performers –

LC: and perversely, Paul who [played the role of] the lift attendant, often they would ignore him, as if he was 'only' a lift attendant. We also had people wandering around looking for their tube train.

DR: These things are all important elements of the whole project of each performance. With this space and certainly with *Tropicana* being the first show in this space, the little, grey door in the station that thousand and thousands of commuters would get a peek into and see something which was quite unexpected, that's part of the shows as well.

LC: Exactly, that door –

JM: into another world –

LC: yeah, *Tropicana* starts with that door and what's behind it, on the other side of it.

DR: With *Contains Violence*, the police block that occurred because so many people phoned in to report strange activity, that's part of the project, all of these things, there's the show and there's also a whole lot of other ripples that surround the show that are part of the art.

JM: Thinking about the ideas behind and within and around the work that you create, specifically in relation to the forms that you choose to communicate those ideas, would you talk about how you shape the work for your audience to experience the concepts at the heart of it.

LC: We all have different views on this. My view is that I expect an audience to make sense of the work and embrace ideas in the same way that they would a room full of abstract art, where they would have a reaction to that and that is what they base their judgement on. For me the fact that it is for some people indecipherable, or that it's very personal or that it means different ideas to different people is fine. If there is an element to 'get' in order for them to 'get' something else then that's different; then we need to find a way to communicate that clearly. I think the fact that the work as a whole is difficult to interpret doesn't mean it's a mess it means that it has a lot of ideas in it and it's not constructed according to traditional narrative.

DR: We never have something to say in its simplest form. We don't have a political agenda that we're following. We don't have an interest in presenting morality. Of course political ideas and morality can be questioned throughout but never answered. We're always trying to create a piece of art that can be appreciated in different ways, that is very much about an audience being in the same space, experiencing something as a collective, which I think is a political event in itself. It's important that there are many different ways and many different journeys that people can find to connect the dots in our work. I think that you can approach our work from an intellectual perspective and find a way through it. They've all been created from an enormous amount of discussion that draws on so many different ideas. Even though the presentation of those ideas might seem like a complete mess or might seem like random events, it can still be unpicked, aside from its other sensory journeys or other aesthetic journeys or the pure surprise and excitement journeys. In *Tropicana* there was one way to enjoy it, the way a couple of school groups enjoyed it, which was by screaming most of the way through it, approaching the show as a fairground ride and it worked like that but it also worked in many different ways, intellectual ways.

LC: That's important, despite its seeming randomness; the fact it can be enjoyed and interpreted in a number of ways is its strength. A lot of traditional theatre, its success is reliant on how successful the writer has been in creating a conventional arc of communication, understanding, emotion in the right place and I just think that we're trying to find another way.

DR: Any element could be the thing that takes an audience member through.

JM: An observation I've made about your work and the way in which others write about it is that corporeal memory is key to the intellectual interpretation. By that I mean intellectual sense is made via those sensual memories we have of it; the way in which we recall it is very much in the body. How do you respond to that?

DR: I suppose a very important thing about making a piece of live performance is that all it can do is exist in the memory. It exists while an audience is in it and then it exists in the memory and it can never be taken off the shelf again. It's very important how it can still continue to live within an audience after it's done. A lot of the work is so memorable in terms of a particular image or a particular moment or a particular surprise that its something that an audience continue to work with for quite a long time afterwards and that's very important because that's all they've got.

LC: That links in with how you interpret it because you can't hang on to the memory of the story –

JM: because it's not a memory that's just in the head, it's a memory that's in the whole body.

LC: That's a very exciting way to think about it, corporeal memory; I hadn't ever thought about it in that way, but yes that's it.

DR: Another interesting thing is the building, the space in which it occurs, you have this experience within this space and then daily you continue to walk past that opening door and you know of the world that lies beyond it and you know some of its secrets inside. I think that's quite a strong thing. It's that Steppenwolf idea, 'magic theatre, not for everyone, madmen only', the idea of a sign you might see on a door that then isn't there.

JM: How do you respond to [Michael] Billington's quote that Shunt 'needs to move beyond sensory titillation and rise to the demands of narrative'?

LC: We're so engaged with that discussion ourselves, these are the things we know and talk about and the implication that we haven't thought about that, or that we're not thinking on those levels, that

Figure 2.2 Image from Shunt's *Amato Saltone* (2006), courtesy of Lizzie Clachan. Reproduced with kind permission of Shunt Theatre Collective. Photo © Shunt

we haven't got the ideas or research or dedication involved in creating something that has substance. [Billington's] searching for a narrative in the way in which he understands, so for him light and sound may be titillation but it's not for me, it's absolutely the creator of art. Clearly if we're trying to communicate something and we fail to do that and the audience get lost – and we can feel that sometimes – then that would be unsuccessful, but he's not talking in those terms. Just to qualify what I said before about wanting the audience to treat it like an abstract painting, I was imagining myself walking into a room and having a really good time, but I wouldn't want to be involved in something that was deliberately bamboozling the audience just to be clever, I want them to get involved with something.

DR: Another reason for that is that our shows are very difficult to categorize because there are so many different elements that make them. Each piece is very different and its intentions are different and the way that an audience move through it and their role in it is always very different. There's no formula. We've never known how any of our work is going to be received because we really don't know what it's going to be. If we were ever confident that something was going to go well then that project probably wouldn't have been worth doing because it would probably have meant that either we or someone else had done it before and that's where the confidence would have come from.

LC: That's why it's important when we make a new show that we're not trying to emulate something; that we are moving on.

Details of the performances referred to can be found at: http://www.shunt.co.uk/

2.3
Akram Khan: The Mathematics of Sensation – the Body as Site/Sight/Cite and Source

Akram Khan: I feel the body is like a museum but an evolving museum so it's constantly mutating. It's a museum because it carries history. It carries generations and generations of information, cultural, educational, religious, political and so on. Then with each generation the body transforms, takes that information and responds to the environment that we live in. For me, everything that you expose to it, particularly as a child, is crucial. That is where the source of most of the interesting material lies, not in the grown up stage. It's really going back to the child body that I'm fascinated by, where the creativity lies. Most of the things that happen to you as a child, the body then shapes it up to deal with in the present.

Josephine Machon: You've talked of 'the world inside the body' and from what you're saying that sounds like it's formed from childhood, yet the dance language you explore is then shaped by the adult experience in terms of the way the dance is communicated.

AK: The first thing that I do is, for the first two hours pretend to put the camera on and I give the dancers a simple task through improvization. I leave for two hours then I come back in and they're still by that point. Maybe some of them who are not so experienced will be still a bit earlier, because they've run out of all the information that they've gathered in the present, very recently. Basically, they use all the technique that they have so roughly about two hours later they've run out of ideas. The body stops. And that's where improvization begins. When that begins, for me it's going back to the childhood body. After those two hours I turn the camera on and things from the childhood come out. For me it was very much the quality of Michael Jackson and break-dancing, as a child it had a huge influence on my body, even the Kathak as a child is very different to the Kathak as an adult. It's much more educated as an adult, much more honed in, more controlled, more refined. As a child it's raw. In a way I try to investigate that rawness. So all the material that we devise from comes out of that half an hour after the two hours of getting rid of all the given stuff.

JM: Following on from that in all of your pieces, certainly latterly, there's always a sense of the bodies on stage working out or working through something as they move. Working through a memory, an experience or an idea; using the dance to try and make sense of that.

AK: Lloyd Newson tells his dancers, before you move, know what you want to say. Anish Kapoor who I'm working with again on the new piece with Juliette, he said, don't have anything to say, just move. For me the best way is when you don't necessarily have anything to say, but you're saying it.

JM: You're discovering what you have to say?

AK: Yes, rather than making a statement you're finding it. In the West, artists regard theatre, music and dance separately. When I see a musician, I see dance, a dancer and I also see an actor. When I see a dancer I see a musician and an actor and vice versa. In a way for me when a dancer moves the first thing I want to *feel* is are they saying something without me telling them what to say, do they speak? And that's very subtle, it's a presence that they have. I remember one particular dancer called Inn Pang [Ooi] who was in *Kaash* and when he auditioned for another company, I was there to help them audition, he did a solo where he stood with his back to us and there was silence, everybody else used music. He was breathing in the experience of the tension that was being created by the silence and the awkwardness and the entire thing was fascinating because he was speaking through his breathing, through his back. He didn't do anything – he literally responded to the atmosphere in the room; and all the dancers are watching and everybody's competing with each other, and he stood there for ten minutes and the breathing just got heavier and heavier and the body changed, transformed. He was speaking unconsciously, trying to say something. That for me is when the body is most interesting. In a way, when I talk about that last half hour of improv, it's when people make mistakes where the loss of inhibition comes. That's the most powerful movement material.

JM: Because of the vulnerability and the rawness that you've referred to?

AK: Yes, it's without that knowledge, the knowledge of others. Instead it's inner knowledge.

JM: There's something primitive about that way that humans communicate, that goes back to the instinctive, back to the animal. Presumably then, because that's the way you are engaging with your dancers, the choreography is invested with that means of communication.

AK: Absolutely. It was never like that before *Ma*. With *Rush*, *Kaash*, I knew what I wanted before I got into the studio and it predominantly came from my body. My body dictated what I wanted so in a way I chose dancers who could replicate me. Then I saw other companies, the generation of Lin Hwai-Min from Cloud Gate Dance, Pina Bausch, Macsek, Anna Therese de Keersmaeker and I thought, I'm seeing a choreographer's vision only. These dancers have stories but they're not telling them because it's already given, they're already told what to do, almost like machines; that's a very crude way of saying it as they're wonderful choreographies. When I saw *Kaash* I saw me and I felt down about it so, in a way, I changed the way I look at things. I enter a space without knowing what I'm going to do and from them comes the material through response. *Bahok* really started by me saying on the first day, you physically look tired and the Chinese dancers saying, yeah we're a bit homesick and I said well it shows in your body. So I said, tell me about your home and we started talking about home and missing home and from there it became *bahok*. In a way, the body is best when it's at its most subconscious.

JM: In terms of the developing aesthetic that you've identified in your work, always at the heart of it is that sensual experience, *Rush* for example explores pure sensual experience, engages with sensation itself, the experiential quality of sensation.

AK: *Rush* and *Kaash* were very physical, very physical sensations. Then I thought, well I'm getting older; I have a raw energy still but there has to be something more. I became fascinated by the sensation of the emotional engagement to the body. The emotional sensation became more narrative so my work became more narrative. I create an aesthetic if you like, well I don't know if I create it but there is a certain aesthetic to the work, and then I destroy it the moment people become familiar with it. I'll try to investigate it a little bit more, the second I get used to it and it becomes a little more formed, I destroy it.

JM: Where there is a similarity in terms of a developing 'aesthetic' that is played with, and it returns to both the Kathak and the de Keersmaeker influence, is in a sense of a mathematical equation that is being worked through. Initially in terms of form alone and, in terms of you talking about the emotional narrative, now it seems that it's the body working through something, working the emotion out like a mathematical equation. Now it's sensation and emotion and live(d) experience being worked out within that.

AK: Mathematics is very important in my life, very important. My grandfather is a genius; he was a very important scientist in India when Bangladesh was part of India. I always believed as a child that maybe I had part of his genius. Then I came to A Level, failed three times and realized that's not the case. But it was nice to imagine that maybe I'm my grandfather's prodigy. I was fascinated by mathematics because of Kathak and North Indian music. I discovered at a very early age, my Biology teacher said, that nature is entirely mathematical, the golden number. For example each stem that comes out of one plant, if you follow the circular pattern of each join, it's a certain number, a golden number and it's always precise. I find that fascinating that nature and spirituality and mathematics are so connected. With *Kaash* it was conscious because I was really investigating the language of movement from my own body and mathematics. Now whatever choices I make are influenced by the mathematics but not consciously.

JM: You've already touched upon the hybridity of dance-drama that exists in your work and the 'confusions' in the choreography. Do you impose those (con)fusions on the dance or do they emerge out of the dance?

AK: Both, it comes out of the dance sometimes and sometimes it comes externally, where I impose it on the dance. When it comes out of the dance it happens usually by mistake, because somebody forgets to move somewhere or we're doing something very complex and unison and they make a mistake so I say keep that now. When it's so mathematically complex, with *Kaash* particularly, people say it's like architecture, but it was simply about waves of energy, war. We'd researched how soldiers have lined up traditionally on battlefields, how they would form different lines, the importance of lines, particularly the front line so that it didn't give way. So we always moved like this or this if there was more attention needed here to defeat this side. That was external.

JM: The bodies are becoming the architecture.

AK: Yes. The body itself is an architecture, although I've never investigated that purely. I used to be so precise mathematically, such as four fingers between the wrist and the chest, four fingers between the chin and the chest, eight fingers between the two large toes, everything was extremely precise. So the architecture [of the dance] can be external, the structure. I remember being fascinated by the energy. What Kathak has is an immense amount of chaos that comes of exploding energy but it's extremely precise, there's an intense clarity about it, so it's ordered chaos. When the body is at that level I'm fascinated by it. There's this cross debate about abandoning technique or people who only show technique, you have two parties. For me it has to be the both. It's very important to have the precision and within that you find the freedom.

JM: To return to the hybridity of your work, the sensual interaction of the body with space, design, sculpture, speech: how do those elements impact upon the dance or do they emerge out of the dance? In particular, you work with sculpture a lot. Is this because there's a quality of movement within sculpture?

AK: Yes, it's alive. If I'm honest, I know nothing about sculpture, I just respond to it. With *Kaash* I just said 'black hole', something that is so powerful it would draw attention away from me so that I can fight it; the five warriors against one warrior. So Anish created this enormous black hole and in the rehearsal it just overtook us. That gave me the challenge, about creating a relationship with it because it's very much alive. Anish is a master at that, genius at it, where it's material but it's living, it's a body, it's moving. Antony [Gormley] also with *Zero Degrees*. I'm very fortunate to work with these artists who really know what they're doing.

JM: So again it's an instinctive response, in the same way that you respond to the bodies of the dancers, it's a sensual engagement.

AK: Always. For me it's very important not to start from an intellectual perspective. It's okay to intellectualize after. My choice is to work instinctively first. I come into a studio, I don't know what I'm doing, maybe I'm feeling tired, or maybe they're tired, so we work on tiredness and we start to build layer by layer a scenario. It's more in the subconscious, a reaction, responding to the present.

JM: In relation to the hybridity of your work, I was interested to note that in the documentary on *bahok* you referred to wanting to start from a neutral point so you turned to theatre. What do you intend by the neutrality of theatre?

AK: Actually I don't mean neutral, I mean it should be from a place of the unknown. If I had started from dance, they have all this baggage, all this history of experience. If I start from something different, that they're not used to, they're discovering themselves for the first time, (re)discovering. I needed for them to abandon the dance, the dancer's body and to investigate the person's body. I filmed them secretly outside the studio; observation is very important for me. I observed them sitting, talking. I remember [Wang Yitong], one of the National Ballet of

China dancers, who was always on the phone with her mother and was always like [he imitates her tone and energy] she was really unhappy. So I asked [Meng] Ning Ning, what's wrong and she said, nothing, she's having a great time, so I said 'and she shows it in this way?! She sounds like she's whining'. But that's her nature, she's always a bit sleepy but when she's working she's working, so I made that into her character and that became the duet. That's about misunderstanding also, because for me it was like she was complaining all the time and then you get to know her and you realize that's her, it's just her musicality. I try to observe them on that level where they don't bring the [dance] baggage in, that's what I mean from a 'neutral' place. I find them neutral when they're out of the studio, out of their discipline. So we started from theatre. Sometimes the will is too strong and the will gets in the way. There has to be a certain sense of letting go in the body. They let go in the breaks, they let go when they're standing outside, but when they're in the space, ready, engaged, there's too much will.

JM: The interplay between the verbal and physical in your work is particularly beautiful. What is the exchange that occurs there? What does the verbal offer that the physical can't? Is there a sense for you of one being able to communicate above and beyond the other in a given moment or is it simply about the play between and the musicality of the two together?

AK: For *Zero* it was the musicality. What was wonderful about *Zero* and Larbi's contribution particularly: my contribution was the story and his contribution was the system of doing it, the rhythm, in sync with each other. That was the revelation for me. It is literally for me where text, music and dance, movement, become one. It's musically designed; you have to do it exactly precise to the original score. We went back to the original video that he filmed of me telling this story for two hours, and we chose 20 minutes of it and the movements were extremely rhythmical, so there's already a music to it and movement to it, taken from me talking naturally, just telling the story to him. That for me was the first time I really believed that this was truly a marriage between those elements. So when I feel the dance can't say it, or the dancers can't say it through the body, then we say it through words. There are things that words can say that movement can't say, that you can say through voice that you can't say through the body. There are some dancers who can say everything through their body and that is what I aim for. I am fascinated by text simply because it's a little bit alien to the dancers; some are natural, some are not, but they don't have that [textual] knowledge of an actor. That's both the blessing and the curse for us because if the dancer has to play a character they don't have the tools to evolve themselves into that character. But if they have to play themselves, it's closer. All the characters on stage are observations of themselves, me observing them. So the text comes out of a situation. I give them a task but they're not sticking to the task so they start talking about it. The text comes out of a necessity, otherwise it shouldn't be used. Everything has to be out of a necessity. There has to be a need for text. Some people throw it in for effect, to create the scene, for me there has to be an integral reason why I use it.

JM: Which goes back to that compulsion, it has to *feel* necessary. What about the collaboration with writers and writing, such as Hanif Kureshi's texts for your shows?

AK: With Hanif he is almost like a dramaturg of text. He comes in and he's with me, like a good friend he's supporting me, and he watches me work with them and then he gives his feedback as a writer. So the words are not his, they're very much the dancers or mine. What he has to offer is crucially important and he helps me edit it, designing the text.

JM: Finding the rhythm that exists within it, shaping it, in the same way that you as choreographer shape the bodies.

AK: Absolutely.

JM: I'd like to move on to something that exists in all your work and especially *bahok*, the exploration of liminal states, the in-betweens that exist within the individual as much as geographical and conceptual in-betweens; that sense of trying to locate self, trying to locate experience and identity. What exists in that space that is so rich for you as a human and as dancer and choreographer?

AK: I can't find a better definition; I would say God exists there. Or something above us, something that you can't say in words. Allah means God in Islam. Originally in Sufism it was a breath, you should never actually say the word, it's just in the breath between the inhalation and the point of exhalation that God exists. Now it's transformed itself, it evolves with each generation but originally it was this purpose, that you could never take God's name because it never forms, it's not concrete. That fascinates me, that's the spiritual part that I'm fascinated by. There has to be one of three things in anybody's work that I see that I feel connected to and if there's all three, it's phenomenal. The first is, 'God', something above, something that lifts you, something spiritual or even an emotion that lifts you, something above what is on the ground. The second is 'angel', the message, the messenger, either there is a message or there is something that is not about 'the message' but it's there without saying it, without pushing it. The third is the human, the direct connection, that I feel that that could be me, to see the humanity in it, the vulnerability, the 'nine'. In Sumerian ten is the number of God and nine is the human because it's imperfect, it can never be ten but it's very close. So for me, if all three are present in a piece then it's great.

JM: So the human is that which is of the earth, the physical –

AK: Yes, the imperfections, the things that go wrong. There's no such thing as perfect human beings and to reveal that is very powerful.

JM: In much of your work there is that sharing of a sense of transcendence or showing the ineffable, the human capacity to know, to have an innate, embodied knowledge of something but to be unable to articulate it. It strikes me that you find those moments in your work often at the point where the human is at its most human as well as when you play with metaphysical imagery. Do those moments arise through your exploration of the imperfections and through that you find the magic moments, the transcendence, or is it merely choreographic play?

AK: It's just reacting instinctively. I don't consciously look for spirituality. I do play on the imagery but I don't consciously think that this is what I want to say [with *bahok*] it just happened that Elaine was dancing with [Andrej Petrovic] and we were talking about Chinese Dragons so that duet wasn't based on Shiva it was based on Chinese gods and mythological characters. We used a little bit of Indian gestures but they were mostly Chinese gestures. The Chinese goddess, Princess, that becomes both monster and human being. These stories were fascinating so we started researching them, I didn't know if we were going to use it or not but I was fascinated by it. Then it's the choice of can this be used or not, half of the images created, I don't know if they're going to be used, half of the things we throw out, so it's very instinctive, how things are chosen, there has to be an inner logic for me.

JM: In *Sacred Monsters*, the point at which the choreography was most powerful for me was the point at which you and Sylvie Guillem came together and created 'the monster' because there was something both intensely human about your interaction as well as beautifully 'otherworldly'.

AK: That was the centre of the show; that was the point of transformation for me. There's always a central point of transformation.

JM: I think that is communicated to your audience because that for me was my point of 'transformation' in terms of my experience of the work; the eloquence of the image and the rhythm of the movement. On that line I understand that you worked with dreams as starting material in *bahok*, which connects with the liminal, in-between states. The power of dream imagery and reconnecting with what has gone on in our subconscious is that it fuses the transcendental with the visceral, the otherworldly with the utterly human. Is that something that you were exploring?

AK: A little bit, yes. We were fascinated by poets. Politicians are the worst for only talking about the present, whereas poets touch all time in the same sentence. They talk about the past, they talk about the present and they talk about the future and great poets have an incredible art of being able to do that. I was reading a lot of poetry for the Juliette Binoche duet, *in-i*, and though I didn't show it to the dancers I think it was subconsciously playing on my mind. We were researching with a physicist and a Sumerian and Sufism historian; these two people shared with us their own research and knowledge, so I became fascinated by time. The present has a certain time and rhythm, the dream has a certain timelessness, a much bigger amount of time because it goes into different universes, so the connection there to memory and the connection to home in terms of memory, memories come from and go back to home, was a very important thing to me. The past, which has happened, the going towards the memory is the in-between to the present. That to me is important – whether we get it in the work I don't know but it is there.

JM: It's also present in the dance and in you in that sense of the ancient dance discipline and traditional technique being fused with an immediate, live(d) language in the present moment.

AK: Different dancers have different abilities. My wife Shanell [Winlock] is very much in the present. What I find fascinating is, as a human being, she's always in the present. She can't be bothered with looking at the past and she can't be bothered about tomorrow. I don't deal with life or performing that way, she's absolutely in the present, all the time. She has a different *time* to the way I dance. Partly because I'm a director, so I sometimes step out. In *Zero Degrees* with Larbi, when I was doing the solo he would step out and observe himself and what's happening, he disengages from himself from within and he would see the work from the outside because we were in the early stages of testing it and I would do the same when he was dancing the solo. So I'm going back to the past, to the future asking all these questions of the work. But Shanell has an ability to just be in the present and the present joins the past and the future.

JM: To bring it back to the way the audience experience the work, the idea of the body as cite, the performance being remembered, referenced within the body in any subsequent recall or analysis. By that I don't simply mean the form of the dance being a sensual experience but the ideas within the dance being communicated on this reciprocal physical level. Do you think about the way in which your audience will experience that work?

AK: Subconsciously I think, it's there, but I try to keep it out because it influences what I do, it becomes about pleasing the audience, what might work for the audience, which becomes political for me, I make political decisions. I've been there and I quickly stopped it. In a way *Ma* is one of the most important pieces out of all the work because it's the one with the most questions still unanswered. It has a lot of issues, a lot of problems, choreographically, dramaturgically, in all respects, in terms of text. It's the work that didn't work for me. But it's very important because that's the one I learnt from most. The ones that worked happened by accident, because you don't know. If I did know then I'd be thinking of the audience. I feel there are some dancers who are very good at showing the idea, and some dancers who are very good at being the idea. The ones who show the idea, I think the audience will never *feel* it, when they *become* the idea, somehow the audience feel it. They believe you, no matter how stupid what you're doing may be, if you truly engage internally, they believe you.

JM: You've talked about the developing aesthetic, from *Rush* through *bahok* to *in-i*, certainly in the latter work there's a sense of you exploring big issues but in a concentrated immediate way through the experience of the individual.

AK: *in-i* is the most universal question, do we dare to love, which is a fundamental issue for every human being. Whatever we do, do we dare to love? Whether it's love for the work or love for another. It's based on very personal stories about love and betrayal; my father's betrayal of me in a mosque; Juliette's betrayal of her partner, both very violent experiences. It's the fact that those are both universal questions and very personal.

JM: Do you feel that you have to work in that way?

AK: No, it's not that I have to; I make a choice from what fascinates you. Artists have a way of putting a mirror on stage, but it's not Big Brother, it's not reality TV, some artists do that but for me it's not about that. It's about crafting the mirror in such a way that you see yourself in a different way, a way that you've not yet seen yourself. When *in-i*, there are two 'I's' there; there's 'I', you the observer; there's 'I', the doer, the one who's sharing something with you. When you see yourself in the other, that's when you become connected. When you think 'that could be me',

122 *(Syn)aesthetics*

Figure 2.3 Sacred Monsters (2006). Akram Khan and Sylvie Guillem. Reproduced with kind permission of Akram Khan Dance Company. Photo credit: Mikki Kunttu

that's the point of engagement because you recognize something in you that is either missing or that is very familiar. At this stage, what's important for me is that connection. The older I get the more the news is bugging me. I used to not care; I was a child so wars, Starbuck's coffee, being politically involved with Israel, all these things I didn't bother with. Now, more and more it's affecting me. To not have knowledge is one thing, to be ignorant about ignorance is a worse thing and somehow that's infiltrating into the work. I'm trying to be very careful about not *telling* what it is; I want to suggest, to leave it open. For me it's what I would say Peter Brook does, because I was hugely influenced by Peter Brook, it's what he calls 'the formless hunch'; a smell or a colour. *Mahabharata* is such a complex work and yet he strips it naked, Krishna becomes an old, white man, whereas actually every character is codified by colour, by hair, every detail is so meticulously designed and yet he strips all that away and goes to the essence of it and that allows the audience to enter, to imagine. The imagination of the audience is an important tool. In that way, most of my work is quite simple in its scenography.

JM: The work is about those imaginative leaps that we take with you. I'd say your work is all about the questions; it's rarely about the answers.

AK: That's a nice way of putting it.

JM: *Zero Degrees* in particular and *bahok*, being left with a sense of questioning, because you're not left with a void but the sense of those questions existing.

AK: Most of the time there is no answer. I mean, do we dare to, what is love? What are your expectations of love, can you love them without receiving something from them? Can you love them simply as they are? So many questions that we're going through for this new work.

JM: It's interesting the way that art explores love; it's often when it's at its most brutal. The work of Sarah Kane, *Cleansed* for instance, the point at which love is shown at its most extreme is the point at which the most extreme, violent, brutal acts are often occurring.

AK: And that's why we've chosen those violent scenes, violent stories for *in-i*. When something comes like adrenalin, a shock, suddenly all five senses are awake, and that's why you remember those moments, that's why the memory is so strong because it remembers all five senses at the same time.

Details of the performances referred to can be found at: http://www.akramkhancompany.net/

2.4
Marisa Carnesky: Trapping the Audience in the Fantasy – Instinct, the Body and the Magic of the Experiential

Marisa Carnesky: I come to things quite instinctively. One thing leads to another, like a jigsaw puzzle or a road map. I've always had a strong idea about what I'm interested in and I make shows about that. I have always had a strong visual and conceptual approach and I've always been drawn to like certain things very passionately. What I've done over the years is put things together in odd orders and called them my shows. So, rather than pinpoint all the various elements that I'm drawn to, such as the carnivalesque, to define each separately, I'm drawn to a series of influences that lead to other influences and it builds like a mushroom cloud. I start with a small palette, so I might be interested in Jewish mythology and I'll mix that with these old dolls and I'll mix that with my interest in stripping at that time, so things come together and they form a fabric. It's not that I want to fit things together than don't fit, or that I want things to fit together hand-in-glove, I want to find an interesting idea that I find exciting. What happens with a good idea, which is usually a fusion of other good ideas, is that it develops and becomes stronger. So the first incarnation may have holes in it and then I refine it, making it coherent, so that it becomes something interesting for people to experience.

Josephine Machon: A strand that runs through your work is this idea of the experiential. All of the influences that can be identified in your work, whether that be carnival or burlesque or illusion or mythology or history and so on, they all engage with an otherworldliness which you allow us to enter.

Figure 2.4 Marisa Carnesky's *Jewess Tattooess* (1999–2001). Photo credit: Manuel Vasson

MC: There are a lot of people who are reviving old circus acts. I'm attracted by traditional entertainment that's come from a history of popular culture, of variety, but I'm not interested in reviving old variety for its own sake. I'm particularly interested in using forms of old variety entertainment, burlesque, magic, cabaret, circus side-show and fairground, using those spectacular and popular forms to explore cultural politics that I'm living through, which is what differentiates my work from practice that just revives the act for the skill itself. What I think is important is where performance traditions meet sociopolitical discussion, like the history of feminist performance and criticism, complicated ideas about the times we live in that fuse politics and sexual identity and meaty subjects, not necessarily emotional subjects but historically and politically engaging stories. I'm interested in looking at those stories and exploring them in a mix of popular entertainment and live art, making it accessible. I like things that have a popular aesthetic but the subjects I deal with are not necessarily popular subjects.

JM: That thinking, those ideas are present in the work often in an embodied manner; you embodying the ideas you're exploring, or presenting those ideas via visual and sensual means.

MC: I want to make work that is emotional and moving. I won't make a show about an abstract emotional state but I'll make a show about facts and history and culture and politics and I want people to engage with it on an emotional level.

JM: And you do work those ideas through in a way that is lived, that you have embodied, like with *Magic War* where you related those political and historical ideas to your own life and practice, the political was made personal.

MC: *Magic War* at Soho was a series of tableaux, fitting stories together, but by the time I'd got that on to tour I'd made it fit better, made it much more clearly about conspiracies. Sometimes my trajectory is difficult for the audience to understand, so I was more brutal with the editing and we added connecting text, direct address to the audience so that they had a road map of why I was connecting things together. It became more readable and easier to understand from that reworking. With each show, I don't spend a lot of time in the rehearsal room, I like to shape it and improve it live. I like to work it in front of the audience.

JM: There's a degree to which you discover what you've got as you work it through, which is an integral part of your process. By that I mean, it's not simply a case of as the run progresses the piece settles and improves as could be the case with any piece of theatre, but there's a shifting quality that is active, that allows the work to become something other whilst maintaining an essential backbone.

MC: Yes. The best way to experience my shows is at the beginning and then at the end.

JM: Something present in your work is the significance of the body in general and your body in particular as the source for ideas, playing out ideas on and through the body, working out conundrums through the body.

MC: That's mainly because I trained not in theatre but in dance. Also live performance has to be about the body because it's 'live'.

JM: Although with *Jewess Tattooess*, *The Girl From Nowhere* and *Ghost Train* in particular you explore the interplay of the live with the mediated and projected body.

MC: More so in the new *Ghost Train* because we now have moving mannequins as well as live performers. I think the body is present in ways that are right for the show. With *Magic War*, rather than it be about aggression onto my body, especially with the history of magic, with a lot of magic performance it's about women having violence enacted upon them as a fetishized act of illusion and I wanted to turn the tables, I wanted to be the one cutting people's heads off, so my body was less involved in that. In an earlier version of *Magic War* I ended up naked with a black bag over my head, on all fours, inviting people to come and cut my head off and it was awful. I just thought, it was interesting in the 90s, it felt radical to get naked and it doesn't now. Now I question why I would be naked, it feels too exposing because I am always representing a version of myself. It's not that I wouldn't use nudity again in a clever way but it did used to be something that I did in very early work without questioning it. I guess the more you work the more you start to question your own practice and with *Magic War* it was about the wider idea of violence to people's bodies, not violence to women's bodies.

JM: For *Jewess Tattooess* your body was a key signifier in that work. It wasn't just a narrative about a fairground Carney who happened to strip; there was an overriding exploration of personal history that connects with wider history. Within that your body was direct working material as source for ideas as much as the canvas for the performance.

MC: It was. It was about the tattooed body, about historical pain that is inherited into your body and the tattoo being an expression of that inherited pain and persecuted identities. In terms of employing your body on stage it's also about what's normal for you and I was working as a stripper at that time and it didn't feel difficult or taboo for me to be naked on stage. Now I find it more risqué and interesting to take risks with my writing and with complicated magical equipment. That is now more interesting for me than taking risks with my body.

JM: Although *Jewess Tattooess* wasn't only about your body when it was naked on stage everything about it was the play of your body. I saw it in its various incarnations, one of the delights being that it did shapeshift yet it always had that strong concept and theme at its root. That being the play of your body in various forms; not just the live body on stage played against the mediated body as you tattoo yourself in close-up, alongside the play of your filmed body, literally turned in on itself to create a variety of other characters in the story. So the intellectual ideas and the written narratives are all played out on, through and around various aspects of your body.

MC: Yes definitely. The whole place I was coming from then, working as a stripper and exploring these cultural ideas around femininity and the body and the Jewish woman's body particularly, alongside the tattooed body being something that the Jewish woman's body is not supposed to be. So it was exploring all these different lines of discourse, culture and history, the tattoos being so connected to the history of Jewish people because of the holocaust, plus the tattoos being bound up with blood, blood being so taboo, specifically women's blood being so taboo in Judaism. It was an exploration of all those elements through the tool which is my Jewish woman's tattooed body. There was a real, clear process. *Ghost Train* came out of that show, from the [filmed] Grandma character with Eastern European, Yiddish folk tales and the travelling show-woman and exhibited tattooed lady, the tattooed body as a funfair side-show exhibition, that got me interested in fairground more and more. *Ghost Train* is not focusing particularly on a Jewish story. In fact, the story in *Ghost Train* has evolved so now it follows a mourning mother looking for her disappeared daughters in an unknown town somewhere in Eastern Europe. In *Jewess Tattooess* my interest in disappeared people is slightly touched on through tattoos marking the body and those Jews who were tattooed and then disappeared. It then follows through in other pieces like *The Girl From Nowhere* or *Magic War*, in the magic illusions where people are made to disappear. Now what the *Ghost Train* will explore, and explain through the mother's opening and closing speeches, is the idea that when people are disappeared there's no closure, they don't get a funeral as there's no body to bury. Again, another element of *Jewess Tattooess*, the final result of being a tattooed Jew is that you're not allowed to be buried as a Jew. So it was very much about the dead body and me mediating upon the fact that to have tattooed myself means I can't have a Jewish funeral, I can't be buried with my family on consecrated Jewish ground and questioning what that means to me; which leads me to the question of do I believe in any of this in any way, because if I do in some little part of me, I've purposefully gone and messed that one up. There was a lot about funerals in it. With *Ghost Train* we've figured out that the central point of it is that it's a living memorial ground where we mark the moment of the people's disappearance in a poetic way; where they can be remembered by an emotive, moving, living moment of their disappearance.

JM: I saw *Ghost Train* is its first form on Brick Lane and the ideas of disappeared bodies permeated the experience, without being spoken. We travelled through this ride where with each revolution, or revelation, a series of bodies in various rooms would be revealed and then penultimately they were no longer there and then finally they were all present together but only as a mirage, an illusion. That sense of present-absence was very strong, fused throughout with fairy-tale imagery and suggestions of sex trafficked women and migration and so on.

MC: It has evolved so that now the entire interior of the train is a series of trains and stations instead of non-specific rooms. There's still lots of ladies in vintage dresses but now we're also looking at women at war so there's nurses and border guards and it's a night train and you're looking at other night trains.

JM: So still it has an essential otherworldliness, which is now literally between borders.

MC: Yes and between time. We're in the night, sleep time, so there's a somnambulist, dreamy feel to the whole show.

JM: An area you've touched upon is your commitment to the writing within your work. Your scripts are very playful and clever. It's in the speech texts that we can feel a lot of the intellectual ideas being playfully constructed and worked out, played out alongside the visual and physical action, illusions and so on.

MC: Like I say, I did actually train as a dancer and I like telling stories visually and I never thought I was a theatre person or a writer. I play with forms intentionally, I'm not a purist I'm more of a showbiz hack. I jump genres at will just to see if I can do them without any proper, formal training in any of them. As regards the importance of writing, I really like to research things, I like a certain rigour in an academic way, I like for those connectors within the piece to have been researched so it's not vague. I do like to underpin my work with that research and I like the writing to reflect that and have some meat in it. What's interesting with the writing is how it appeals to different audience members. I like it to have that cross appeal, the historical rigour fused with fresh, hip ideas.

JM: It also exists in the visual and physical imagery. There's an interchange between the writing and the visual, both underpinned by this embodied thinking.

MC: Yes there is. On this *Ghost Train* I've been working a lot with the designers; I come up with a series of sketches and I ask them to help me develop those into working designs for the builders. So we wrote the story visually first and as we came up with the visual story and named each character we then came up with a back story for other characters. Generally for me the writing of the text does come from visual ideas or bold concepts.

JM: To return to the idea that you take us into these other worlds, the 'spectacle aesthetic' in your work, you enjoy playing in those other worlds and also creating an experience for your audience. This experiential quality, you taking us on a sensual journey such as leading us along the streets of the East End with *Carnesky's Burlesque Ghost Box* through to the sensual journey of *Ghost Train*.

MC: I guess I am interested in a dreamy old, funny vaudeville show as an aesthetic and you get a lot of that in Eastern European and European independent film and theatre. I wonder why I like spectacles, I hate to bring it back to some Jewish thing but I was brought up with a really strong sense of the holocaust, my dad is obsessed by it, and if you do see these images of piles of bodies, it is a spectacle of the macabre, it's of epic proportions. I am interested in big, bold, symbolic things, which is kind of cultural. If you're Catholic or Muslim or Jewish you've been brought up with large symbolic meanings, reading stories about rivers running with blood, it sticks in your mind. I love a bit of ritual. Everything I loved as a teenager about popular culture was bands like Siouxie and the Banshees. I do genre hop and I do make strange shows that explore current experiences but I have to do it through a lens of dreamy, nostalgic otherworldliness because that's what make me feel moved. And if my audience want to come with me then they're very welcome to.

JM: So your work is a fusion of genres, a fusion of the sensual and the cerebral.

MC: Yes it is sensual and cerebral. There's a lot of circusey, imaginative, dreamy work out there, like James Thiérrée of the Chaplin dynasty, somebody in a tattered dress falling out of a wardrobe for about an hour. And I love that, I'm in the dream with them but if they could just tell a story about politics or something then that's what I'd be really interested in and that's what I'm trying to do. I like lots of things and I just put them all together to try and find an imaginative language. When it works it's been a bloody complicated journey to get there because I've tried to shove everything in and it's like a mathematical equation; how do you put all this stuff in and get one thing out that says all that stuff. Like with *Ghost Train* now, it's really dense, there's tons of stuff in it but there's one resonating point to it that branches out to many other points. It has a dense cultural mix of things, because I really like work like that. My favourite film is Jodorowsky's *Santa Sangre*, it's a really dense film with a lot going on, brightly coloured and dreamy at the same time and I'm trying to make stuff like that. I often think of shows in filmic terms, lots of my references are not theatre, they're films and I'm working out how to explore those ideas as a live experience, as experiential, so it's like being in the movie. I want the audience to be trapped in the fantasy to be engaged in the other world, which other companies like Punchdrunk and Shunt are also doing and possibly that's because we're all from a film generation.

Details of the performances referred to can be found at: http://www.carnesky.com/ and http://carneskysghosttrain.com/

2.5
Naomi Wallace and Kwame Kwei-Armah: Desire, the Body and Transgressive Acts of Playwriting – on Writing and Directing *Things of Dry Hours*

Josephine Machon: Naomi would you begin by clarifying what you intend by 'transgressive acts of playwriting'?

Naomi Wallace: I guess it sounds kind of grandiose; one has to think of what one is transgressing, so for me as a political writer it's about looking at social relations, often through a very personal prism. The transgression is how do we write for theatre in a way that challenges not only mainstream ideology, and I mean mainstream ideology in the United States and Britain, but that troubles what we are taught, what we are told, what we are given about sexuality, about race, about class; I know I'm being very general here. For me theatre, because it is live and history is enacted through the body, social relationships are enacted through the body, is the perfect medium in terms of looking at resistance through art. I don't even want to say through art because that sounds like you're separating it, art and resistance. For me theatre comes alive when it's about issues of power – who has it who doesn't, who tries to get it, who's taken it away, all enacted through the body.

JM: So playwriting as transgression in terms of philosophical concepts and content. What about playwriting as transgression in terms of form?

Kwame Kwei-Armah: What's interesting about *Things of Dry Hours* in that way is that it attempts and does, very successfully in my opinion, transgress on three levels at the same time; form, in terms of the poetry, in terms of the richness and the denseness of the poetry that allows us to enter the beauty of language and the power that it holds; it also subverts notions of race within the arrangement of the characters and the way that the story is; and it then also looks at gender, so that at one moment in one play there are these three approaches hitting you, stretching you, transgressing at the same time. When I read Naomi's essay about transgression it confirmed how wonderful it is when people speak about a philosophy that you actually first encountered through their art-form, as opposed to solely through their prose. That's what happened for me with *Things of Dry Hours* and all of Naomi's work.

JM: Why make political ideas sensual, so that they are *felt* by the audience, so that intangible concepts are encountered physically and in that way, made tangible? Why do you feel you have to communicate those philosophical, political ideas that way?

NW: For me it's twofold, and not separated either; one is to do with myself, as a human being moving through my little piece of history, my little piece of life; and then it has to do with my interaction with the social world, with history. On one level it's about exploring humanity whilst maintaining my own in this economy, in this society and that has to do with transgression but also exploring the idea of what is our value, how do we resist being diminished because of our gender, because of our race, because of our class. I'm very interested in modes of resistance that are not often obvious; sometimes desire is a mode of resistance. So for me that is the site where these things can come together through actors collaborating with directors, it's a living art-form and many have spoken about it so much better than I can but because the body is the site where power is enacted or power is struggled over, then the power of putting that body on stage. And of course I create a history for it but you also have the body of the actor acting that history. I write poetry too but it seems to come alive for me in three-dimensional space in a way that it hasn't in writing poetry or fiction.

JM: I think that's the question I'm asking; why write the politics as poetry, in that sensual, embodied way, as opposed to the David Mamet or David Hare approach which is starker, cuts straight to the intellectual 'telling' of the situation?

NW: When you sit down and you listen to people talk, when people are uttering who they are or trying to utter their experience, their story, or speaking about their family or what they've suffered through, it's usually far better, more poetic, than anything I could dream of. Those pieces get inside you and you reformulate it. Yes there's the Mamet style of speaking, but that's also a fabrication, just as mine is in a way. I hear that poetry when people speak, what about you?

KKA: Yes. Interestingly enough, our first encounter was on a page of *Time Out* together, talking about August Wilson, whose speeches are terribly poetic. In terms of language, there is isn't anything that's mutually exclusive but what attracts me about hearing the poetry of my community, for want of a better term, and trying to lay that down on the page, which is what I automatically recognized in Naomi's work, is that it serves the community from which it comes; because invariably what it says is that, you too have the ability to articulate at the highest manner – truthfully – not so that someone can say that's a wonderful scan but through the images. Naomi also writes from a class perspective and who more than the working classes across the world speak in metaphors and similes and in wonderful stories to educate. I don't find it an act of transgression I find it a very aggressive form of honouring and being loyal to the essence of those who struggle and find themselves within the dispossessed.

JM: From what you've both said there's that sense of this language being lived; it comes from and goes to the visceral so the receiver can only *experience* it as a result because it conveys that live(d) history in the live moment.

NW: Right. Just as an example, something that really moved me, I was reading an interview with a young man in Gaza and he said, 'there is nothing left of us, we are only shapes' and I just thought, I couldn't write something like that. It's not that I would take that but there's a way of listening and, like you're saying Kwame, of giving back, of allowing that articulation to come from those who haven't been heard to articulate their own lives, their own agency. Yes I reformulate but there's a way of speaking, a way of telling stories, a way of articulating oneself, one's experience that has an inherent transgression in it because those voices are oppressed and not heard. We're not just talking about some art of witness here; we're talking about characters on stage creating their history and social moment, in that moment.

JM: If we could go back to something you've both touched upon, the idea of the body as the key signifier, the body that speaks for itself, that is as eloquent as the words it speaks. In particular the idea of the body as a resistant force, often at the point that it is most oppressed, for example Roach on the chair in *Slaughter City*, the body fighting back –

NW: or I was thinking of *Things of Dry Hours* where Cali plays with the shoes. In the production that Kwame and I did, there was this wonderful strange play and sensuality that was set up with those shoes. She would move around these shoes in this kind of dance which was quite dark but it began with play, like a child and it went to this other place. Yes it was an object she was using, a daily object that we need to survive but she transformed that and in doing so, in that moment transformed herself into something else, something she wasn't even comfortable with.

KKA: You're absolutely right; one of Naomi's most magnificent skills is to be able to present the body as the final frontier, the last thing to be broken; the wonderful nakedness in *Things of Dry Hours*, the elementary form. I feel there's a great spirituality in that because the body is also the place, much like prayer, where divinity and humanity meet and struggle. That was so evident when I read this. Where Cali starts putting porridge on her face and the boot polish on Corbin, this moment where the body becomes the manifestation of the psychosis, then immediately moves on from the psychosis to become the manifestation of aspiration and then moves on from that to become the manifestation of horror for the audience.

NW: I like what you're saying, because in that way the body is not static in any way, it's in continual movement, continual transformation, the body itself and the signs it takes on.

JM: That's a perfect illustration of where the play with the body in your work fuses the Brechtian and the Artaudian, the political and the metaphysical; demonstrative, gestic acting that moves into the unusual, the transcendental hieroglyphs that burn to sear an image, an experience into the body of the perceiver.

KKA: Yes. The merging of the political and the spiritual. It's a lesson for me as a playwright in how you fuse the two without aiming for that, rather aiming simply to serve and fusing that through a combination of spirit and intellect.

JM: Something else that's also present in your work in relation to the body is the sex; an exploration of both sexuality and the corporeality of the body.

KKA: Naomi is rooted in her femininity, which is rooted in her intellect. The first time I read *Things of Dry Hours*, as with the first time I met Naomi, you get the impression she's done lot of internal work making sure any rubbish that has been put in has been processed and thrown out to create this oneness, which exists in the work. There is an absolute, strong sexual tone, an energy. The men are incredibly sexual with each other, shaving, holding knives to the throat, walking around each other, talking about their members. Some writers are good at writing sex but not rooted, human desire; that desire that's elemental, that will strip you to your core.

JM: What I'm interested in is the idea of, in the same moment, the small, immediate, personal stories become the vast, social political, historical stories, both played out on and through the body. How is that related to this rooted desire that we are confronted with? How does the social, political and historical connect with the sexual?

NW: Desire has its own intellect. It doesn't run parallel with our minds. It's like Cali, she's drawn to Corbin but at the same time she sees where he's broken and blind but she's still drawn to him and she finds out her own things about herself through that, through moving towards him. I was thinking about the changes that we made to the play; now the sexuality between them happens in the play and they both have agency and there's negotiation between them and it's much sexier for it.

KKA: We often, particularly when we go within the realm of the intellect, relegate desire to something beneath or hidden, or we exploit desire to say that this is something other. What Naomi does in her work is to absolutely synthesize the heart, the mind and the body. That thing that drives everything we do; I walk into a room and I listen to someone that I'm attracted to just that little bit more than someone I'm not attracted to; I see a man in an outfit that is well put together and I listen just that little bit more because I think, subliminally, there's something that he can teach me. That's what Naomi does with her work for me, to not negate the trinity, the body, mind and spirit.

JM: Absolutely, those moments of intense physicality are the kernel of the work. Do they not become a paradigm for those moments of transformation? That point at which the desire is worked through or made manifest is a moment of transition, of positive change.

NW: I was just thinking about how much censorship there is to our desire, aside from the age, gender, race and class stuff that comes in; it's almost as though our own desire moves ahead of us and we have to catch up with it in a way. Of course there's that conflict between, as with Cali, what she wants and what she knows but, and I guess this is the transgressive part of it, she finds a way to make those come together, to allow herself to move forward with him. It's partly to do with her educating him and that culminates in his love song to her, which is talking about dialectical materialism but has to do with kissing her and her body. He tries to do it through language but it's also worked out in physical ways.

KKA: That was one of my favourite sections; to use dialectical materialism as foreplay. It was a masterstroke. In my opinion it's what the overwhelming majority of articulate men do; they use their language as foreplay. To put that in the context of breaking down sexual barriers and addressing centuries of sexual inequalities and power; to put that into this white male at that period, using that language to a black female to say 'let me have what I want from you, I will use whatever I need to, my best tools, to get you'. There is moment-by-moment interplay between the actors having to follow each syllable whilst also guarding themselves from them. For Cali trying to shield herself from every word and yet at the same time make him aware that she knows what he's doing.

JM: And it also says that political thought, political action *is* desire and allows desire to come to fruition.

KKA: I was fortunate enough to grow up at a time when afro-centricity and politicization was deemed sexy. What it meant was I could chat women up by just talking about the ancient kings and queens of Africa and our contribution to world civilization; that was my foreplay! In this piece I saw that wonderful integration of a wider politic and a sexual politic, the fusion of the sexual and the politic.

NW: What Cali doesn't allow in that scene is for him to kiss her. What was important to me about that is that the modes of desire that we have been given are rejected by her in that moment. He's wants to seduce her with that language and then kiss her and she reformulates the energy and desire between them by saying, no, you're going to do it the way I want to; these are the places that I'm going to allow you to touch me and these places have stories. Not only are they physical places on her body but also her history is written on her body. That is sexy too; as opposed to there's my history written over there and my body's here. Her history is in her body and she teaches him that so he moves to a place that is not the mode of desire granted him but to a place of listening and learning from her as opposed to taking the lead, so to speak. Again, he's looking for the way to seduce her and he realizes the only way he's going to get close to her is to follow her. That's a different place for them to be in and a different power relationship to that he is used to.

JM: Staying with this idea of experiences and ideas being played around, through, in and on the body, in your work speech takes on a visceral-verbal quality. Speech is reclaimed as a physical act, not just in the act of saying but the experience of the words becomes corporeal.

KW: It's not often that you get verbose characters who are not vainly verbose and not only articulating and giving insight to the motivation of the character but who are educating us at the same time. To talk about the power of those speeches is quite difficult; when we speak, poetically or not, when we are passionate about what we are saying we either over-enunciate or we *move* because the words are not coming out of our mouths as quickly as they're forming in our bodies. What happened in our production was that these speeches were coming out of, like fire. All I was saying to the actors was, every word is a new word, every word is just landing and you're trying to get it out to make way for the next three that are piling up. So with these big speeches we watched the audience follow every line, go with it, because not only were they wonderful stories but also each syllable is written out of a desire to communicate an idea *and* an emotion. Often dialogue is created to give either an idea or an emotion. The duality of idea and emotion or passion is particularly evident in the speeches of *Things of Dry Hours*. As a director it was hard not to over-dramatize because just from reading the prologue you could hear this orchestra of words, the rhythm is so beautifully set up. It's because it comes from the fusion of the spirit, intellect and body that you go to those places.

NW: There are different ways of performing my work and at times I liked a more detached, more formal, presentational approach and Kwame did something completely different. What I saw happen was the language, the stories; everything became embodied in the actual bodies of those characters on stage. Kwame's a phenomenal writer but this was his first directing project, and I loved the way he talked about my work. It was his brain and the way in which he talked about the play that made me think I don't know if he's a good director or not but I want this experience. So in terms of embodying the politics of the story and the desire in the body, I think the changes that Kwame suggested for me were toward fusing these things. When you're using the body, the mind and the spirit, there's the danger of melodrama, especially with the heightened language. Maybe I was even afraid of putting Corbin and Cali together, allowing that to be in the moment, fully alive, as opposed to where the man's dying. Those changes allowed something else to happen. It's hard to know what some of our own censorships are, fears that send our work in a certain direction.

KKA: Putting my playwright's hat on, one of the hardest things I did was to work up the courage to say to a writer who you really admire, actually I think that you're not fulfilling your own brief. We still didn't really get it until the first night; we were still playing with it in terms of me trying to realize what you'd written on stage.

NW: The changes Kwame suggested allowed the play to be more dangerous, in its desire, in its moments of power between Cali and Corbin. For a lot of reasons I hadn't been able to get there and I remember Kwame being so gentle but being very clear and, I said to him at the time, I've had suggestions made about my work before and how I know if they're right is if I feel afraid because I'm not sure I'm up to doing what's called for.

KKA: Something that's very important to me, as a black writer, and this isn't about blowing smoke up her arse, but she was the first white writer to have captured a black voice, and captured it not out of a syntax but out of an absolute truth. It was an understanding of more than the words, because there's a trap you can fall into which is to think, well this is alien to me so I'll write it verbatim or you can write it Standard English so that 'anybody' can speak it. What Naomi did was a wonderful combination of heightened language, of colloquialisms and of truth. How brave she was to tackle this character, putting him at the centre of this narrative and putting language in his mouth that was true to him and yet true to us all. It's a human syntax that is really hard to achieve.

JM: What is opened out for you by writing from an alternate perspective? What does it allow you to encounter to write from an alternative perspective?

NW: On a basic level, that is what's 'done'. We don't always write from a different race but we write from another sex, from a different gender, a different age, from different generations. On a very general level, I'm interested in writing about American history and without African Americans at the centre of American history you're not writing about American history, so that's on a political level. On a personal level, in the five years it took me to research and write it, I was looking at what it is to be white in America and the privilege of whiteness and the damage of accepting the myths of whiteness. I'm going to generalize here but white people have lost their humanity for not recognizing their privilege; for not looking at their part in a racist system; for not taking responsibility for their part in that; for not resisting that and not because you're doing something nice for a community but because it's damaging to your own humanity. DuBois has written about this, James Baldwin has written about this, in beautiful ways about the damage that racism does, not only to the people that it is enacted upon but also to the perpetrators, to white people. I didn't want this to be a play about working through whiteness. This was about a man who was the centre of the communist movement in Alabama, he was the centre of it and that's what I'd wanted the play to be but how I'd written the end was not fully getting there. I laugh with Kwame about this and I can joke with myself about it, my whiteness, what I've grown up with, which is still a very racist culture in America, was pushing the white guy to the centre. I can laugh about it but we are not naturally writing from our souls. It's like we have to rip and tear and throw things away to get to anything that might resemble real truth. So the garbage of what I'd grown up with was getting in my way. Kwame had said, yes Corbin is in there, but the central story is the father-daughter story, what they teach each other and what they come to learn. The struggle is when you decide what kind of theatre you want to write. Who do you want to write for?

KKA: You've got to write to express the melody that's suppressed within us and hope that melody will become catching.

NW: Corbin does enter the scene because I was also interested in, and one has to be very careful about this, especially with the abominable history of white writers and how they have portrayed people who are not white on the stage, looking at the interconnectedness of our history and the race struggle in these moments of intimacy between black and white people; where were those moments where they could stand together? One of my favourite books about his whole things is *Another Country* by James Baldwin. He has this astonishing love story where a black woman and a white man are together and the white guy wants to be closer to her and she's telling him there are places he will never go or understand. What he doesn't get is that by acknowledging that, by saying, 'you're right there are places I can never go', that's how he can get closer to her. Not by the specific knowledge but by acknowledging the difference of experience. I think I tried to work some of that out between Cali and Corbin and Tice. There is a friendship between these two men, even though Corbin betrays him. It's not sentimental in that way, even if it's only here and in this way they connected, the fact that that existed at all is the possibility for us and for the future, even if it passed, even if it failed.

KKA: And you can love. This is a play filled with love; everyone falls in love with the other person on stage in some way, at some point. The biggest love story is Corbin's love story, Corbin falling in love with Tice. He falls in desire, in want with Cali but he falls in love with Tice.

There's a moment where I wanted Steven to just look at Tice in that way where you go, 'damn-fuck you're fine'. That moment where someone walks into the room and you just miss a heartbeat; and that's in the writing. There's a moment where Corbin just looks at Tice and goes 'you are more than I ever expected – you know me and can teach me and I am learning and I've got this gig to do but you can make me better'. Everywhere he looks in that room there are people who can make him better. I wonder how many marriages are based on that, the, 'I love you because you can make me better, you can take me to places I cannot go by myself.' That's at the core of this piece, that recognition of need, want and love. I know it because I have that with a few people.

NW: That skip a beat thing, even if you can say desire has its own intellect, there is something that is inarticulate about desire. It's that skipped beat. There's a space there and we're talking in metaphor to a degree, but if you skipped it there is a new space there and that is where possibility is. In that moment when someone literally takes your breath away and I'm not just talking about, 'oh I wanna fuck that person' but everything about them stirs you up, puts you upside down, that's when I say even desire in itself can become a form of resistance. What is the time period, what is that space, what are the social politics of that time? That still influences that space, even if that space is inarticulate and is driven by desire. That fascinates me. What dynamics are set up in that moment for that skip to be able to happen?

KKA: This is going off tangentially but I can see that with America and Obama, that space, that absolute, 'oh my god', it's that space where it's held for a moment, where the change happens, where you negotiate with that change.

NW: You're exactly right and whether one agrees with all of Obama's approach, there's something happening in our desire for Obama that goes beyond Obama and it has to do with the possibility for change. It transcends him but he is the vessel for it. That desire for him, they can laugh and call it Obamamania, but it's an intellectual desire, it's a desire for this articulate, amazing figure who's so different from what we've had, all that coming together in that one moment, that skipped beat.

KKA: I had it with Obama, swear to God, when I heard the 2004 Democratic speech and in fact that influenced my choosing to do *Things of Dry Hours*. It was the merging of the political and the possibility of change and I went back and listened to Obama's speech, I had it on a podcast. Can you imagine what Tice would have thought about Obama?

NW: It has to do with what Lefebvre said about when history or possibility breaks through the present moment, there's this space, even if it closes or contracts later, and if you present that on stage, what it means is if it happened then it can happen again. We can widen that gap, that beat, one hopes.

JM: Which brings me to this timelessness that exists in your work – very much played out through Tice in the Prologue and the Epilogue, in *Things of Dry Hours* – you create that 'beat', that space within those bodies, within their moments in these plays. Through them you draw attention to the potential.

NW: I was just thinking about the apple at the end, about Tice as the deliverer of this message, he reads it and the last image we have is of the apple but we don't get to know what the words are. There's something in what you're talking about in the image of the apple, his expression, 'well I'll be damned' and the beat that isn't articulated but you know it's there and it's been read, so it's released.

JM: A moment of transcendental imagery. We've been talking about the corporeality and the very physical nature of your plays but for me they also contain a transcendental quality, which supports this idea of the fusion of the mind, body and spirit in one and the same moment. Naomi, you've talked about imaginative acts being key to the characters' potential for transformation. Something that's key is the fact that the imagination of the audience is key to their experience of this work, to the way that we make sense of the work and enter those moments. It allows us to engage with the ineffable. That exists in the final apple image, we don't quite know what is there but we can fill that space if we want to or we can allow it to be about the not knowing *being* potential and where that might take us.

KKA: To just add to that, the metaphor that most moved me is Cali and the sheets, that for me is the quintessential moment of Naomi magic.

NW: I do know that in the material, whether it's a cup or a body, what we are taught that we can do with our bodies or with an object that can be pushed to a different space so that it's re-envisioned, like her sheets. These are a real material representation, as well as a real thing, of her oppression and through her imagination whether she's sleeping or dreaming, the force of her imagination takes this object and creates something of freedom for her, that she can dream on, that she can play with, that she can go to another space, physically, emotionally, spiritually with. It's hard to talk about those things, you can see them and you can feel them work but why they work is the question. It has to do with resistance too. No matter how we are diminished or defined, there are moments where we're able to take ourselves out of that and see ourselves in a different light, in a free life. You see that light in Cali, you see that come out of her, it's the first time we see her laugh.

KKA: What's wonderful about that moment, what was profound for me was that it was the moment in the play when the power of the mind is investigated in several different forms; articulated in a poetic form, controlled through the body. This was a non-verbal moment of mind-power, an absolute celebration and a challenge to those that watched to believe that this can be. That there are moments of inner freedom, whether this is in her mind or whether it's actually happening here and now, there is a place that one can get to where the power of the mind is so concentrated, is so pure, is so strong, that you can lift yourself from your environment and transform it. We often hear those stories from men who are in prison who say I know my body is here within these four walls but my spirit is not. And that's what that moment is, that moment of asking what is the true nature of human potential, and that is what is at the core of this play. What we can achieve if freed, if free.

2.6
Linda Bassett: Bypassing the Logical – Performing Churchill's *Far Away*

JM: Caryl Churchill's work is visceral. It's (syn)aesthetic in that it affects the audience in a sensate manner and engages the intellect in unusual ways.

LB: Bypassing the more logical brain you mean?

JM: Bringing it in at a later point, or at least, allowing the 'illogical' to have the importance of the logical. Certainly not bypassing the logical, but engaging it in a different manner.

LB: Because I often find with Caryl's plays that when people try and follow them with their head, it doesn't work – they get stuck. What I mean by bypassing it, if you sit there and let her work on you, you get the experience. But people who fight it and try and make sense of it along the way get stuck, and then say this doesn't make sense.

JM: I feel *Far Away* is a return to the visceral in verbal language, to allowing language to bypass the logical. Following on from her plays that were exploring a sensual interplay with all those different languages, dance, design, music such as *The Skriker* –

LB: Which, I think, *Far Away* is a direct descendent of –

JM: whereas *Far Away* is highly concentrated.

LB: It's like a homeopathic remedy, it's been distilled and distilled and distilled so that you say one word and it's doing the work of four sentences. That's why it felt like we were doing a three-act play and it only lasted 45 minutes, because she's pared away everything else. No, I think distilling is a better word than paring away because I think that's what she's done, distilled it down to something essential. She never changed anything – she took out a couple of sentences. Often with a new play things are being changed all the time – but no, nothing. And that's not because she's precious or anything, it was just because there was nothing to change. She's done all that work already.

JM: I'd like you to talk about your initial response to the play, when you first read it.

LB: For me, that's the purest moment. In my opinion, it has to be savoured. It's your first encounter. It's the last time you have a chance to get what the audience might get later on. Because, from thereafter you're going to get all sorts of stuff going on, but that first reading is you meeting the play. If it doesn't get me then I don't want the job.

JM: What do you mean by 'get me'?

LB: Find an echo in me I suppose. There are two things. That it puts out something that I want there to be in the world and also that it asks something of me that I think I can give. I remember I did make quite a lot of attacks on the play while we rehearsed it, which must have been very hard for Caryl. You have to unleash your negative feelings which I find a very necessary step, because otherwise it's all just holy, holy and I can't bear it. But that was very hard with Caryl sitting there. I remember a very difficult rehearsal, partly I found it difficult because I find Harper so destructive and I think Caryl and I have a basic disagreement about that. She's much more forgiving of Harper than I am. I think it depends whether you see Harper as a human, or an archetype. I wanted to play both. And as an archetype she's unforgivable but as a human being she's completely forgivable. We spent a lot of time trying to find redemption in that last scene. I think it's interesting that *The Skriker* does the same thing; because I saw that as an audience member, not an insider. Right at the point of transformation she finishes the play and she's done the same thing again with this. I'm really looking forward to the play she writes where she gets to the point of transformation and transforms instead of ending it.

JM: Isn't a strength of Churchill's plays the fact that you do have to go away and work through that?

LB: Maybe that's true, maybe that's the glory of her work. I don't know how I feel about that. You can still carry away something that's gone through a method of transformation. When I saw *The Skriker*, which was a different experience because I was watching it and at that point of transformation that's when I started to get really interested, that's what I mean. I think that's what some people may have found with *Far Away*.

JM: Because they want 'the answer' to be revealed?

LB: No, it's not that. I think it's, 'this is now exciting, I knew the rest, the rest was interesting but I know that – but this is really exciting, we're getting down to the goal now, the really difficult stuff'. That's what it felt like for me.

JM: The final speech is one of those moments, whether or not Joan steps into the river is revelatory in that sense, it articulates something inarticulable.

LB: Well she's a poet, and she writes poetry, it's a poem more than it's a play. That whole last scene, with all the animals and everything, is mind-blowing, to use an old hippy phrase, it's like taking a mind-expanding drug. Because that's what she's doing, she's expanding your brain. People are tempted to say, well that's silly, cats don't kill people, or that speech, 'they were killed by ...', and so on, but actually all the things she says do kill, heroin and –

JM: coffee

LB: pins were in there – [laughing] yes I did find pins difficult.

JM: grass –

LB: Oh the grass tried to remain neutral – they were 'burning the grass that wouldn't serve', that was the bit that would regularly make me weep.

JM: Why?

LB: Well with the horror of it, the horror of what we do. That everything has to join the war. It's apocalyptic. We used to suffer, we all suffered. The way we clung together – that's why when we went to The Albery we chose to share that dressing room because it was like we were little people up against this horror. We needed the human warmth of each other just to keep going. I think doing it twice a night did take it out of us actually. It was like doing Greek Tragedy. It didn't give you anything as an actor. Some plays as you do them, you're giving something to the audience and they give something back to you. Other plays, like when I did Sophocles, you don't get anything, it's all out, you get nothing as an actor, it's all give and they get it. So you can come off feeling like complete shit, and the first time that happened to me I thought, well how can that be? It's because it's not two way; it's the same with *Far Away*. People would say, oh, I'm so excited, my brain's excited, and I say, well don't you find it too bleak and they'd say, well the subject matter's bleak but the form of expression is exciting, so it's not a depressing experience watching it, it's enlivening. That's what you're talking about isn't it, the power of the language. And that's why the audience don't despair.

JM: Kathy Tozer said that the rehearsals of the final scene led to the most revelatory moment. Kathy started to cry and it was all bound up with so many complex things, but on a simple level, if Joan can't even come home and talk to her husband, then what else is there and Churchill said, I now know what this play is about, I didn't know what I'd written before. How do you remember that rehearsal?

LB: Well I remember, when Kathy cried, thinking, thank god, this is where it has to go. And I remember Caryl saying, her eyes getting moist and saying, I'm going to cry now, which I think was extraordinary. She said it as if there was something wrong about crying. Which I thought was very revealing about Harper, and the whole thing about the play, that feelings are not supposed to be expressed. It's like the water – at last the water was there. We'd found the redemption.

JM: What was the redemption that you all found?

LB: That *feeling* hadn't been destroyed, exactly what you were saying, that Joan isn't dead. She's still alive. The sad thing is that in the world of the play, it's too late. Or maybe it isn't. If people can still feel then maybe it isn't all lost. I think why it's important – the tears, the feeling of it – is that that's the only thing that's going to save the world – feeling. People having their feelings and not going against them. That's what stops you –

JM: but what about feelings of anger and hate, they can generate that situation.

LB: It depends what you do with them, you can feel angry, doesn't mean you have to hit someone over the head. It gives you your sense of value is what I mean. How you *feel* about something is different, from how you *think* about it, you have to think as well. We do an awful lot of thinking in our society, feeling we're not so good at. That's why that seems to be a redemptive thing, if Joan can still feel. And then you could even look at the statement 'it's such a shame to burn the hats with the bodies' that something's actually coming up from the back, that if she can feel that, then she might be able to see that her feeling's misplaced, to feel for the hat, and because she brings up the bodies, it might even be that she's feeling for the bodies.

JM: In that respect, that final scene describes in essence, (syn)aesthetic performance. When you say 'bypassing the logical', that's what it's arguing for, for privileging feelings. That's what that scene is arguing for.

LB: Yes that's interesting, I think that's what Caryl's – asking for, vying for, inviting people to do.

JM: Both 'making-sense' and *'sense*-making'. Following that, I'd really like to talk about the exploration of language, the physicality of the language, explored in rehearsal; the movement in the verbal language, because there was very little physical movement.

LB: Apart from that Act Two, where they worked out that ballet, that very technical business. But certainly Act One and Three were very still. I think that you're affected by just saying it. And sometimes you don't get it until you're performing it. That's why the French call rehearsal *répétition*, because that's what you're doing, you're saying it over and over again. We didn't sit and talk about the nature of the language, we said the words and talked about how we felt about them and whether we didn't understand them, you know, 'what does she mean when she says'. For the first scene, Stephen [Daldry] did a thing where we drew pictures of it. So that Annabelle [Seymour-Julen, as Younger Joan] drew what she'd seen, then I drew above it what I'd seen. We took a long time to get hers and then mine spun off from it and mine got more and more ridiculous, more and more difficult to draw. I found that a really useful rehearsal because I felt quite exposed standing by the wall with my felt-tip trying to draw the story.

JM: Drawing the lies.

LB: Yes, because they keep changing. I remember when I got to the end, drawing the heavens, and your soul and the stars, I got her to come and do it with me. She joined in with that one and that felt significant, that I'd got her to join in, to contribute to my picture, when my picture was actually a lie. She didn't scribble it out, she made it more interesting.

JM: It's interesting that Kathy talks of *Far Away* as 'sparse' and you said, 'distilled'.

LB: I thought of that on the way here, that that's how I experienced it. They say about homeopathic remedies that because they've been distilled, they're more potent than a large amount of drugs. I think that that's true of [Churchill's] work, that it's more potent than other people's bludgeoning of writing.

JM: In relation to that, I'd like to go back to what you said about this being the natural progression from *The Skriker*. What do you mean by that?

LB: [Laughing] I don't know what I mean.

JM: Because I agree with you. I think it's something to do with the language being so disturbing, being damaged, fairy-tale damaged. In *Blue Heart*, for example, the language is entirely damaged, but it's deconstructive. By that I mean it's primarily intellectual and playful rather than dangerous, sensual and emotional as a route to the intellectual.

LB: *Blue Heart* marries *Skriker* and you get *Far Away*. *Skriker* has so much content, no that's not fair on *Blue Heart*. I just experienced *Skriker* so much as form *as* content – I think that's what you're saying – I never penetrated the content of *Blue Heart*. All I know is as a poet, she's utterly true to herself, and you can't say that of everybody, and that's what makes her great. Caryl felt she had to defend Harper against me. Getting the human side of a character like that is fairly simple, it's in the writing. It's getting the evil side that I found hard. When I finally got the crocodile speech it was a breakthrough for me – to be able to condemn crocodiles like that because I, Linda as a person, do not anthropomorphize animals. So, condemning crocodiles, I mean I know she's not just talking about crocodiles, she's talking about people, but she is, she's using those words. Harper's talking about real crocodiles but it's a racist speech. So of course it's brilliant that she's saying it about crocodiles because what she's saying is disgusting, if she ever said it about other humans, it would be appalling to listen to.

JM: I think that adds to the complexities of that speech though, because it does also suggest that we as a species think that we're above animals, we have the right to destroy other species, as well as other races. That is appalling.

LB: I was surprised how I found it, I mean, I've played Medea who killed her children but night by night I found it appalling. The thing about my struggle with the crocodile speech, what Caryl wanted me to do to make her human, I think it was the opposite. I think what I'm saying is, it would have been so easy for any actress to play the human side of Harper, what was hard was playing the archetypal side, because we're not archetypes, we're people. So I only got the crocodile speech when I really did finally get that much hatred going, because it allows the audience to feel the hatred. There's a certain part of you where you're trying to lay it bare, not underline it or be obvious about it but just let people feel the power of hatred. In order to let them feel it you've got to mean it. If I didn't invest it with the racism behind it, I don't think it would have had the power it had.

JM: It's interesting you talk about the brevity. When I first saw it, my friend and I couldn't speak for a while after it and we both said it was like holding our breath throughout. Coincidentally, Kathy said she felt that Caryl had to hold her breath to write that last speech. I think the whole piece is like holding your breath, which comes back to your wanting transformation, is it the exhalation that we're wanting?

LB: It's interesting how the beginning came about, the singing. There we were in the technical rehearsal and we were up in the little studio, and you know what technical are like, they go on forever, and the curtain was down, they were talking about lighting, and I was sitting at the back on my own, and I was going into that yoga thing where you put your head on the ground, The Child I think it's called, so just sort of resting, and I started singing the song to myself. It's a song that I learnt myself as a kid. I grew up in the fifties, but there were still a lot of war songs about, we used to go around chanting about Hitler even though none of us were even born then, leftover songs. One of them we used to sing, like a school dinner song; 'There is a happy land, far, far away, where we have bread and cheese three times a day. Egg and bacon we don't see, they put sawdust in our tea, that is why we gradually, fade, fade away'. It's based on a hymn, there's a hymn about a happy land. So I'm singing this just because of the words and from in front of the curtain came Stephen's voice saying, 'Linda, could you just sing that again', and I sang it again and then that was it. And then Caryl said later that she thought of putting that in the play, because she knew the song, slightly different words, so it had obviously been in her head. She must have found it very peculiar sitting in front of the curtain suddenly hearing a disembodied voice, singing the song that she'd thought about. And I loved the way that began. I wanted Harper to be alone there for a minute, on her own, maybe it was the storytelling, the Grimms' fairy-tale aspect of it, that Ian [MacNeil] had put into the set as well. Maybe it was to do with the feeling of the child being, is she real, because there's a moment at the beginning that you don't know whether she's a literal thing, so it allowed there to be that, she's part of Harper's brain. I used to go through all sorts of fantasies in my head about it, you know, feeling lost, Harper having witnessed knowing what Uncle was doing outside, so somehow summoning the child to ask her those questions, because she couldn't bear what was happening in the shed.

JM: The child is the materialization of the human side of Harper?

LB: Yes, they're part of the same psyche. Not that Harper's a human being and Joan isn't. There all human beings in the play, but in the wider sense, it felt good that there was that moment where you didn't know whether or not the child was a real child. All the people that were gathered to do this production seemed to be in agreement that there was this other aspect to it. I would be interested to see a production where they were far more literal and to see what that did. You could do it in all sorts of different ways.

JM: Churchill's plays do have this open quality that allows the real to blend with the magical. It's that that makes those intangible things tangible, makes you *feel* those ideas. She presents abstract ideas on a familiar level so that you can connect with them on an unfamiliar level. It's that that makes the writing itself, the performing of the writing, a sensate, visceral form.

LB: I just though of another aspect, when we were rehearsing, and I remember saying this one day to Caryl in rehearsal, and I remember feeling it when we did *Fen*, although it was slightly different because we'd done the workshop process, that even though Caryl isn't dictatorial about the way you say her lines, you have to find your way how to say them, often there's only one way to say them; and Caryl will very simply say, 'have you tried stressing that?'. It's like those lateral thinking games you play as a student. I always find her work a bit like that, you know those things like, there's a man in the room and he's hanging from a beam, and you'd find these fascinating journeys that your brain went on, the conclusion itself was often rather dull, but the journey, because you'd go to wild places, and her plays remind me of that, they take your brain to places. When I saw *A Mouthful of Birds* the first production in Birmingham, the first night, and I stayed there for a week and I watched it every night. I didn't understand it and I didn't even care, I didn't want to. And as the week went on I understood more and more. But the first time I saw it where it was hard to follow, was the best. It was fantastic. And as it got clearer, I felt I had a less rich experience.

JM: So what was it about the more difficult experience?

LB: Well it's what you're talking about, it does that to your brain –

JM: disturbance –

LB: yes, you can't go to the easy, the familiar palaces in your brain. It's almost like you can feel your nerves, your brain cells being forced into a different synapse than they want to go, and it's very exciting. I think she stands out from her contemporaries, like Hare, she stands out, and she is a woman, but it's not because she's a woman, I think it's because they're more literal, more clumsy, and she's scarily deft. She's not clumsy at all –

JM: everything's so precise –

LB: fucking laser precision. But can you see why I'm longing for her to do the next bit, because if she can do all that, then if she wrote her true feelings down, I think it would be her masterpiece. Something amazing. And I'm not saying that what's she's already done isn't amazing, but there's something waiting to happen there for me.

Details of the performance referred to can be found at: http://www.royalcourttheatre.com/archive_detail.asp?play=247

2.7
Jo McInnes: A Text That Demands to Be Played With – Performing Kane's *4.48 Psychosis*

Jo McInnes: What was difficult at first, when you get that text written like, not in book form, lots of sheets of white paper with words on, you're so shocked that you're actually doing it as a play. It's very difficult to disassociate myself now from what we ended up with, and how easy and simple it seems in a way. To start with the main drive that we all had was to understand as much as we could. We cut it up into sections and we titled each section. The first two weeks we just workshopped it, just played with it.

Josephine Machon: That's what I'm interested in, how you played with it – because it is a text that demands to be played with.

JMc: We tried to sort of base it somewhere, to try and find the reality of it, with the doctor and patient scenes really, to try and find the truth of them, and then we'd swap over, and we kept playing with different versions of that. The actual mixed voice pieces where we all spoke, we played with those in very different ways. Sometimes James [Macdonald] would give us a situation, like the one that happens right at the beginning, 'Body and soul can never be married', that section, he'd say, maybe, you couldn't sleep, or you've an urge to release something, you're in your bedsit on your own, and then we'd all be doing this at the same time but you wouldn't actually say anything until you felt the need to. What's great about James is to begin with he never gave you very strong guidelines; he'd gently tease it out of you. It's difficult to remember now exactly what those exercises were because an exercise would come from an exercise; it was all born from what we were doing, so there was no structure really. Apart from getting to know the words, just becoming familiar with them and trying them in different ways, playing off each other.

JM: Following on from that, did all the exercises tend to be language based?

JMc: We all became very physical, things like 'flash flicker', there was lots of self-indulgent, embarrassing afternoons spent trying to work through that, that was the hardest thing. We just used to try and find what they might mean so we each would have a go at trying to create a meaning for them, and it would be off the cuff, you'd never think about it, you'd just do it, and you'd all show each other.

JM: Did it become more dance-like in that respect?

JMc: Well it sort of nearly went there and we all went 'no, what are we doing!'. I mean, there's a great place for all that but to us it didn't seem true to the text at all. It seemed to bounce right out of it, you know. Keeping it in the reality of it all the time, because the words are so extraordinary, I think it felt to us that to pin it down in a reality of any kind helped us to believe it.

JM: Because the words themselves are flying and so physical –

JMc: Yeah, because the words themselves are so expressive and physical and strong, to then add physical movement is too much, you're over explaining, it becomes self-indulgent, it becomes about you and not about the work. Later, we'd each have a go individually on each section, and we'd each watch each other's, so you'd play it with your own understanding in mind. We did quite a lot of that which was great. We all knew all of it, there was no splitting up going on until right towards the end. There was one section that really helped me when I was going home on the train from here, 'Avoiding My Heart', which is what we called, 'I dread the loss of her I've never touched'. I was coming home one night on the train and everyone else was just doing what they were doing, it really helped the connection with, you know when you're in a very crowded place but you're absolutely alone, you're absolutely cut off, and the reality of that place helped me to understand this better and to give it a drive. So the next day when we were reworking it, I played it with that in mind and James really liked that, and that's where the video of people walking across [projected on the white floor of the playing space], that's where that came from. One person would spark off a little idea, which someone else would take on, that was happening all the time. It was the only truly creative process that I've ever, ever been involved in, and I've done 11 years of theatre now. Although I've really enjoyed, and I've learnt so much from some of the other plays I've done, I think this was the truest in a creative sense for me, just because there were no strict rules and no strict guidelines. Even directors who say they don't have any do. James is very open and we really did work with each other, we really did trust each other. We really did allow ourselves to be vulnerable in front of each other, extraordinarily vulnerable. And our work, and I know that the other two [Madeleine Potter and Daniel Evans] would agree with this, our work came from that, we fed each other. The thing about us speaking in unison, that happened naturally. We understood each other's rhythm and thought so intuitively. We all knew all of the play. Character comes from that I think, character came from those journeys that all three of us took together. Little pieces you would pick out, and then slowly but surely a character developed.

JM: I wondered if you had tried to explore a sense of character in order to make sense of it for yourselves. Did you ever try to think it through in terms of character, the patient/doctor roles for example, did you have a clear idea of 'character' in your heads?

JMc: No, and we felt that was really important, it was very important for the doctor in particular. The woman in the play is definitely a real person, obviously, to us all. That whole play is her. But the doctor got very, very naff and shit if we started 'doctor acting' it because, before you know it you fall into this doctor role, and that's where the final disconnection came from [the performers never held direct eye contact with each other throughout the performance], because although we'd played the doctor/patient themes, for quite a while, facing each other, we found that if you ever looked at each other, it broke the concentration. As soon as there was eye contact, something else happened, we don't know why. Especially with the doctor/patient thing, because suddenly a whole other play walked in the door. Aside from the naff 'doctor acting', the relationship changes, the tense of the relationship changes. We always saw those doctor scenes as being memory, even if they were immediate, they were always a memory, a memory of something that had happened. If you start to do it another way, you bring it right into the present, it places it and it makes it too real, it gives it a realistic edge that isn't helpful. That was one of the ideas that James did come with. He really thought any face to face communication, any realism in the scenes was going to throw everything off track, we had to find another way to make them true. We couldn't rely on conventional ways of doing.

JM: That's a feature that's really subtle, but is part of the impact of it all feeling like a dream. That you were tracing something, rather than it 'really' being there. Following on from the role-playing and the 'disconnected' communication, can you tell me about the workshop on the text that you participated in prior to being cast, where there were a number of performers involved. I'm interested in how the players had been reduced to three in your production.

JMc: Sarah's notes on one of the first drafts of this play, did have three characters, Victim, Perpetrator, Bystander. But then she scrapped all that. I think, two women and one man, the oddness of three, I don't know, exactly. To me, the odd three, two women and a man, just makes perfect sense and I can't say why, but also it would make perfect sense to do it with ten or sixteen. When James saw us on the first day of that workshop he told us that it was for him to work out if The Royal Court were going to do it, and to understand how many people would do it. I think three just adds so many dimensions, because you've got two people playing the scene and the other person still in the scene.

JM: You've talked of reading the notes that Kane wrote on *4.48 Psychosis*, did she provide any reason for why she wanted to open it up and no longer be three characters.

JMc: I think she just wanted to keep getting away from form. She just wanted to keep shifting form, breaking form. I don't know, maybe she was never going to write those characters, they might have just been a tool for her to help her write it. I didn't want to look at those notes actually.

JM: Because you felt it invasive, like reading a diary, or because she wasn't there to offer it?

JMc: Yeah, it was a bit strange, but then I thought, if she wrote them and she trusted Simon [Kane] and James with them, I didn't have a problem with it, but I was happy enough not to hear them.

JM: Do you think that from a performer's perspective you didn't want to hear them because it might suggest the way in which it should be done?

JMc: Yeah, and I think Sarah doesn't want any of that, she deliberately doesn't give you any of that. She deliberately makes you work hard, and if you don't work hard she shows you up to be the wanker that you are. I think with Kane's work more than any other work, the audience are the fourth actor, they are the other person that's involved because it takes their involvement. There was one awful night, it was so quiet you could hear a pin drop, and afterwards we thought we'd done alright, we were blaming it on them and James put us right, he told us that we didn't let them in, we were so dark that no one was allowed into it. It was too dark, too heavy, too serious. And that was our penultimate show.

I learnt so much every time we did it but every time we started again I was as nervous as when we did it the first time. Always terrifying, not once I started, but the preamble to it, the warm up, to find that dark place where you can sit so comfortably in that dark place that you can be funny, you can be flippant, you can do all those things to give the play the texture that it deserves. But if you're not in that place to begin with, you're too light or too dark, then everything feeds in on itself. We were moaned at the other way as well, that we'd made it too light, too easy, too accessible. It's a really difficult line to play. [Kane] demands a lot of you as an actor, but you have to be in the right place. If you are in the right place it's amazing to play, sounds corny but it really does trip off your tongue. Nights when we were all really connected, it was the most – that last scene, I actually felt that I levitated, quite a few times. I was actually floating, and I've never had that experience before. And all three of us got it. We'd often experience the same things at the same time. Daniel and myself, at exactly the same moment thought someone had a gun, and was going to shoot at the stage. For us to be feeling exactly the same thing at exactly the same time, the connection of it made our hair stand on end just thinking about it –

JM: It had a transcendental quality –

JMc: completely, and that's in there. That's how good the writing is, it takes you to places that you don't know. That is what is so good about the play, that it does have a spiritual, whatever the words is, a supernatural, spiritual, huge place to it that you do really visit. And we visited it together often.

JM: I'd like to talk about the 'fuck you' speech, which you performed, because, although it's much shorter, much more concentrated, I feel it resonates in the same way as A's love speech in *Crave*.

JMc: Well, I know the love speech in *Crave* was written about someone, whereas the 'fuck you speech' is more a realization that that person doesn't exist. I suppose with all the work I've really enjoyed I've connected it with love. I played Sonya in *Uncle Vanya* and that's all about the loss of love, the need for love, and to find that same truth in [Kane's] play suddenly opened it all out for me.

JM: You mention Chekhov, did you find you made connections with Kane's work and other writers?

JMc: Not a specific work, but I found lots of connections – Shakespeare, because of the way that the tragedy goes on, the form of it, the language, the control of language definitely, the shape of it in the way that you can learn it, be part of it and then be very imaginative with it. Sarah has that same gift. It's actually very, very structured, and the punctuation is very strict. But, it's to find that structure and then find how creative and imaginative that structure can be. The language can be as structured as it likes really, it's your commitment to it. Once you've committed to the rules, you really can have such a fantastic time, and that's what I mean about it tripping off the tongue.

JM: You said in our previous conversation that *4.48* is 'still very much in me', I'm interested in that notion, of it still being in your mind, in your body, your flesh. Can you explain further what you meant by that?

JMc: Because it means so much, it's so true We're all sick, we're all ill, we're all fucked up and if we say we're not we're lying. Culturally we're all absolutely rotten, where we're going, what we're doing. *4.48* strikes out at all that. It's not about a single journey, it's about all of our journeys, it's about how we all live, the bullshit we all swallow. It's about how hard it is to try and be a better person. The way we live and the way we want to be are in such conflict. To me it means everything, it's me talking here with you, it's going on the tube, it's reading about the Twin Towers. There's so much reality in it without it being a documentary, without it representing an extreme realism, it allows people to have their own take on it, it's expressive and free enough not to pin thoughts down. That's why it means so much to me. It's been a massive learning curve for me as a person. Learning how true you have to be in your work, about the depth of truth. Working on something like *4.48*, you have to completely invest everything that you possibly can into it. We'd drink in the bar afterwards and you could see members of the audience thinking we were really shallow because we were elated, really elated and laughing, really laughing, because we could see the ridiculous in everything. When you say those words so surely, that's when you become wise, that's when you see the wisdom. To be able to share in some of that wisdom, I found that – I hate the word proud, I've got a real aversion to being proud but I am proud of *4.48*, I am proud that I was in that play, and I am proud that I was asked to do that play. It helped me, as a person, so much. It helped me understand things, often when I'm in situations now I think of the play. Psychologically it's been like a therapy for me. You've gone through this massive journey and you're still being encouraged to laugh all the time. The last line of the play is *so* positive, 'Please open the curtains', it's just like, all I'm asking is for you to just look around and see what's happening, you've got to be involved. That's why people used to walk out, because they weren't up for that. I don't think that they're really offended, I think that the play strikes them in a way I can't explain. People don't want to be challenged. Our whole lives are fed to us so that we're not challenged. Everything we do. It's very hard to be challenged about the way you live your life and not become extremely defensive. And that what Kane's good at, she's good at challenging. I think plays like *4.48* come by you very rarely. It's such a skill to be able to put truth into language because language is often very evasive, to use language like Sarah does, where she makes it work for her.

JM: How do you respond to the idea of *4.48 Psychosis* being a suicide note?

JMc: To me it's not a suicide note at all – I detest that train of thought. You can't remove it from Kane but there will be a time when people are reading that play and they won't know that she's died. She's writing it on a much longer term than we can foresee. We've been in a lifetime when she died but that play will last a lot longer than all of us and it won't be perceived in the same way. I think it's the antithesis of a suicide note in a way. You don't end a suicide note with 'please open the curtains'.

Jo McInnes performed in the original 2000 production at the Royal Court and in the later revival in 2001. Details of the performances referred to can be found at: http://www.royalcourttheatre.com/archive_detail.asp?play=245 (2000) and http://www.royalcourttheatre.com/archive_detail.asp?play=79 (2001)

2.8
Graeae's Jenny Sealey and Playwright Glyn Cannon: Seeing Words and (Dis)Comfort Zones – the Fusion of Bodies, Text and Technology in *On Blindness*

Josephine Machon: Graeae have an ethos of 'multi-sensory theatre'. Jenny, could you explain what that means for you and what your approach to creating work is?

Jenny Sealey: I think all human beings do not access everything because there's so much information everywhere, all the time. If you access something 'differently' like I have to lip-read, I use sign language; if I haven't got my hearing aid on my voice goes off. Because I'm very deaf people think of me as someone who is less intelligent, less emotional, less of a 'real' person, therefore she doesn't buy into the theatre in the same way as everyone else. Also my previous work before Graeae I was working with other people with physical and sensory impairments creating operas for them, doing Shakespeare with them with people saying 'why are you doing it with them, they don't understand it?'. How do you know? I'm just looking at stripping away the levels of that text and finding whole new layers of how to communicate something. When I hear say a high-pitched scream I think 'she's got it, she's got it', or it may be in the way someone follows the yellow fabric of a velvet dress. Who's to say, who's to dictate what we do and what we do not 'get'? I suppose that whole thing of finding different emotional agendas or different ways of displaying emotional agendas, which is true to the actor, through the actor's physical or sensory impairment that all informs the process.

JM: Do you think accessing the process in that way taps into something that is heightened, more human?

JS: Well it's very real and in one way it's one off because there's more non-disabled people in the world, so we have a uniqueness, and that's great. It's tapping into that. It's about how we hear information or how we don't hear information, for example with *On Blindness*, that whole thing of if you don't see, every single sound matters, every single sound translates into something so you become aware that breath is hugely accessible in theatre. That was something we used very much in *On Blindness* and something I used a lot in *Blasted*. In that production breath was tangible. We all breathe but we forget to hear it, we forget to see it. So that relates to that whole sensual exposure of what any *one* thing is. We forget about that.

JM: Glyn, would you talk about how much you were aware of that approach, how it may have influenced you, at any stage in the process?

Glyn Cannon: I knew Graeae and I knew Jenny's work, I think I must have seen *Peeling* while I was writing *On Blindness*, so I knew about the possibilities and the idea of playing around with the senses. There are two parts; on one level there's a colossal mistake in *On Blindness* in that the central device that I've written is visual, so that seems quite perverse, especially when you know that you have an audience that is largely visually impaired. Then the flip side of that is that there was a lot of visual information in the text, almost too much, and that's what came out so to be honest, the clearer way of receiving it was to not be looking at it. I think I was influenced more in relation to the collaboration, talking with Jenny and everyone else involved. We talked about audio-description and including that in the text but also the big thing that we talked about all the time, a thing that I talk about a lot in theatre in general, is the gap between audience and performance. You want the audience to want to come and fill that in; in terms of information given and in terms of the blanks. That was the ethos to some extent; that there was an equality of chaos.

JS: Also it's about you as a writer and we as directors sort of premeditating some of those gaps and talking about how they might be filled by various different audience members but we were always constantly surprised that they weren't filled in the way we assumed they might be.

JM: Which brings me to my next observation. Both on the page and in performance it feels to me that *On Blindness* explores perception, communication and miscommunication, the spoken and unspoken, understanding 'truths' through getting it wrong. Would you talk about that in relation to form; the tensions between the written and the spoken and between words and movement? What was discovered via finding ways of exploring physically what Glyn had written?

JS: I think, in one sense it's a very naturalistic play, it's very simple and I don't mean that negatively at all. It's got a very concrete through-line but what it does is challenge you to look at your perception of yourself, it challenges you to ask of yourself, if you were going to be painted nude how does that make you feel. The difference between male reactions to nude issues in comparison with female is vastly different and how that impacted on the team of actors. The main character, Edward, one thing that was sort of there but sort of wasn't there with him was blindness as a metaphor for everything, which is dangerous because it can get a bit, you know [JS pulls a grimace].

JM: By that do you mean clichéd?

JS: Yes. But there *are* people like Edward who are so confounded by anything that's different from them; they are stuck, so fucking stuck. He couldn't see his hand in front of his nose and yet his heart pounded for this beautiful woman but she was so outside of his comfort zone that he builds bricks in front of her to stop himself communicating with her because he dare not believe that she can be sexual or beautiful or have desires or be absolutely fine about a boy wanking in front of her. She's fine with it all, she's way ahead of the game than he is.

JM: It's interesting that you use the term 'comfort zone' because it felt like the performance was wanting to play with that as a central concept. In terms of making the concept tangible, it played with the idea of *playing* with people's comfort zones. Is that something that you, Jenny, think about when creating work for an audience?

JS: Graeae always takes people out of their comfort zone. *Blasted* was a prime example. David [Toole], a soldier with no legs, 'oh that's to shock', no he's just a good actor. But the discomfort was not necessarily about the actors, although it was about the *physicality* of the actors. Jenny [Jay] couldn't walk up the steps, she physically couldn't do it but I designed it also to stop her from doing it. I played the stage to her status, she'd get as far as she could and then she'd start to slide back down. Gerard [McDermott] was very carefully mapped out, same with any set for blind actors we know that as the artistic team, but for the audience, a lot of them didn't realize that Gerard was blind, the character in *Blasted* becomes blind but we played a twist on that. The discomfort is the audience's scripted narrative, you see every single one of Sarah Kane's words so the soldier revealing his amputation, fucking with his leg but telling what he's doing, you could close your eyes, although if you're blind you can't, but you've still got to listen so there's no way to escape. He's telling you what he's doing, which is exactly what Sarah Kane wants, she doesn't want you to escape. Design is very important. With *On Blindness* the design concept was just absolutely genius. You describe something and a blind person has to keep it in their head. So you've got a nice set, everything naturalistic and in rehearsal with a few people, it's all there. But what Julian [Crouch] did was draw the set, describe it, then take it away so we all had to keep it in our mind's eye. I think that whole thing about 'the mind's eye' is very prevalent in *On Blindness* for sighted and non-sighted people. It was one of those beautiful moments where it just came together, a real diversity of voices saying the same thing.

JM: Jenny's referred in the past to engaging the imagination and looking beyond 'what one sees' and that informs the Graeae aesthetic, and I just wondered if that, in relation to form, was present in your 'mind's eye' as you were writing Glyn.

GC: Sort of, yeah. It moved a little bit away from what I think I started with, not in a bad way it just did. Initially, I was talking about the two overlapping spaces – this is where Frantic [Assembly] came in – I was very much interested in choreography and movement around and the audience's enjoyment of that. So I was thinking all the time about who was where. I had my ideas about who was where [in relation to a split level/split dialogue feature in the play] and whom they were stood next to but that got developed through rehearsal anyway. So there was a lot about space and the audience having to work hard to keep the realities apart but also enjoy the fact that they're close and so it was a question of space and respecting the audience, asking the audience to juggle.

JS: But it was also about a complete lack of space, that claustrophobia and there's nowhere for all these emotions to go, so sometimes it was 'like [with a tone of discomfort] 'oh I wish this thing would hurry up and finish', you know what I mean.

JM: Can you talk further about this idea of tapping into the imagination, which for me is the power of theatre, making your audience perceive something that is intangible, enabling them to understand the concepts at work in the play when they haven't necessarily been explained in a narrative fashion. Then, coming out of that, this appears to be key to the way Graeae work which is about encouraging new ways of 'seeing', encouraging new ways of doing in performance and forcing the audience to go with you.

JS: I wonder whether part of my meddling with form is a backlash against, for example, when I describe something and sighted people say 'why are you describing it when I can see it?' and of course not everyone can see it. In that description I'm making a decision about how to describe something which may not be the way other people see it. By me making that decision about how to 'describe' something, it is my decision as an artist to say this is how I think it is but it is for your sighted, non-sighted and hearing audience to disagree with that. It's interesting, when I sign audio-description and hard of hearing people say, out of that frustration at being 'told' something, 'but I can see it'. Well, you should see even more.

GC: It's a game, I think for some audience members, once you get used to that style, once you understand the game of it and the enjoyment you can get from that. That's why some people reacted really strongly to *On Blindness* and were against it but it's about letting the audience into the game. With *Peeling*, Lisa [Hammond] would turn around and read the audio-description to see what her next line was. I know she was doing it deliberately but that game of letting the audience into that is great.

JS: It's letting the audience know that we know that the game's there. Thinking about that game though, not every play lends itself to this experimentation. With *Bent*, the play was written nearly thirty years ago about life before then, and it's not our experience, thank god, so therefore the concept in my head was that we are all visiting it as a piece of really important history and when you revisit something you have to have that sense of story telling. So all the way through it was marked with, for example someone would say, 'a captain arrived in SS uniform', they described what was happening to move the story along as actors but in role and it was really fascinating. It didn't always work but when it did it was fantastic.

JM: So there exists in the interdisciplinary approach that Graeae take an uncovering of the sensual possibilities in the work via its very form. In terms of your writing Glyn and the concept of the spoken and the unspoken, it felt for me that, to quote Eugenio Barba, you were playing out this 'struggle against the fixity of words'.

GC: To write in a way that undoes when there's such a tradition of very sacred writing in new writing everywhere. For me to say I'm more interested in play and to try to write less and say that the actors will do this – the literary bit is what gets me. It's not literature –

JM: It has to be performed –

GC: Yeah but so much of writing is about the exact poetry of what you say. The play I wrote after *On Blindness*, *The Kiss*, is so empty, repetition that's played again and again. It's about the emptiness and what's played in the bodies. And that was at work in *On Blindness*. There's a bit in *On Blindness* that I couldn't write until it was happening, the bit where Maria's being described by Edward. But I didn't know that Scott [Graham] and Karina [Jones] were necessarily going to be doing those parts – so I couldn't write the description, well I could have done, but that happened in rehearsals. I did a lot of writing in rehearsals.

JM: That's interesting because Steve Hoggett has talked about how David [Sands] opened up a new physicality for the choreography by working through the signing. As practitioners you became aware of new vocabularies and different approaches that refer to the sensual.

GC: Also the embodiment was a big lesson for me. I remember the Charles Spencer review revealed an interesting point; something that was his prejudice also came up with other friends who were not as bigoted but sort of said the same thing in a much more sophisticated way, they had a real problem with David [Sands] – do you remember one of the first questions David had in rehearsal was is his character deaf – and I think we said we don't know, because it didn't matter to us. That was the interesting leap, because that was the leap for the audience to make as well, to say that they didn't care either – because a lot of the prejudice came around that character who is quite clearly actually deaf because he's signing, whereas there's no reference to that in the text. Yet there was another scene between Edward and Greg where Edward signed back.

JS: That reviewer had stereotyped the characters because of their disability –

GC: and it's interesting the people I've talked to, really clever people, who still have this problem and they got very confused.

JS: One of the things I always try to bring up, because I don't understand it; you have a play and there's a disabled character, a wheelchair user in it, the play's not about disability, it's never mentioned but at the end of the play the audience are clapping and the actor gets out of the chair and bows and the clapping increases – it's like 'wow, a non-disabled actor playing a crip, fabulous' – do the same scenario but you've got a real-life wheelie, so they don't get up, they take their bow twice from the chair and the clapping remains the same. Why?

GC: Although that was an interesting thing though, because the play was explicitly dealing with blindness, so there was one character who was blind, in fact we decided that Maria was more visually impaired than Karina [who played Maria] was. And that became very important to me, it came out of a pub conversation with Simon and Jenny, about how these characters are depicted, and I was very keen that there was never a speech in the play that began, 'the thing about being blind is'. I was really adamant there mustn't be anything like that. I *was* writing that by using the metaphorical thing but nonetheless that was the weird confusion for people because the cross casting activated all these prejudices.

JS: But that's what I think is fantastic about it, and that is absolutely what I think is the way to go. It fucks with people's minds but I really don't care. It's asking questions, provocative questions. I was talking to a casting director and he couldn't find the right disability for this particular role and I said, 'I tell you what, why don't you cast that role as a non-disabled actor but cast the others as disabled, there's equality, that's great' – actor is actor, they can play disabled they can play non-disabled. And he was like [feigned horror] 'no I can't do that' – but it really is the way to go.

JM: There's still something interesting in there to do with audience perception and imagination and playing with that. I'd like to move it back to the sensual, specifically in relation to signing, which is such an exciting performance language, how does the signing open out, perhaps rediscover, what exists in the words?

GC: It spoilt all my jokes [laughs heartily].

JS: There are always some things that do not translate.

GC: No it's because the punch-line has to go first, you start with the gag.

JS: It's funny that, you get deaf laughter then you get hearing laughter, a mixture, fantastic, why not? It is difficult for a writer though – we should play with that even more. We should hold a workshop with writers and comedians to explore how to write comedy so that deaf and hearing audience can react at the same time. But back to your question, when we were working with sign language, we had the luxury of working with the writer in rehearsals and looking at the translation, it's always interesting for anyone who likes words and likes how grammar is worked, grammatical structure, sometimes the writer can see that [translated] sentence structure says everything and I've added on 20 words and I don't actually need them. Or it highlights where the gaps are, where we do need 20 more words. It informs the process. Certainly for actors it really informs it because if they say 'oh I'm going to say this bit really dramatically', in which case, if David's translating, let him know because he'll have to do really big signs because it changes the translation, it could be sad, or heartbroken or a different emotion and there are completely different signs for all those things. It constantly disciplines the actors.

GC: And the absolute luxury in that very first scene where we had Steven signing for Scott's character and David signing for Jo McInnes' character and the movement where everyone was moving around, it was beautiful.

JS: Yeah, we just messed about with form something chronic really. It's always about that; it's what we've tried to do with *Static*, trying not to make signing 'BSL interpreted'. We did get away with that more so with *On Blindness*, with David's character being very voyeuristic, so he was just always there. I think we could have pushed that more maybe, but it never felt like it was direct translation. I think we've made some mistakes with *Static* where we needed a bit more backing translation, there's an imbalance between English for hearing people and non-hearing people. But with *On Blindness* it was truthful – we had it all the way through, we were quite maverick with it.

GC: Making Steven do that much as well, he had to learn it, which meant he had quite a specific accent, he was quite stilted.

JS: So he had to do four or five different choreographic steps, with four or five different choreographic things with his hands. I kept trying to find a language to talk to Steven about how best to break down this barrier where he was just getting a block about it all.

JM: I've mentioned how the signing allowed Steven to access a new quality in movement. Similarly I'm intrigued to know how the signing opened out the text. Those of us that use written or spoken language every day, we can limit it, almost in the same way as Jenny referred to earlier, the 'I can see it so you don't need to describe it for me' attitude – that is, I can't see it for seeing it; we lose the multiple layers that can exist in the most simple words.

GC: I was always really fascinated by the way, with BSL, the inflexion is the beautiful thing. I really love the fact that you can inflect it in so many ways and the inflexion is so personal and that inflexion gives a performative edge.

Figure 2.5 Graeae's *On Blindness* written by Glyn Cannon. L-R: Steven Hoggett, Jo McInnes, Scott Graham, David Sands. Image © Graeae. Reproduced with kind permission

JS: There's a difference of honesty in it, in a way. There's a whole other argument about sign language which is can you have subtext? And you can but it's harder, it goes back to what you were saying about the punch-line coming first.

GC: The subtext is in the interaction of things still, it's pure drama. As a writer it stops you being vain, it stops you bullshitting. An awful lot of writers bullshit and it exposes that. That shook me a lot, I'm still figuring out how that works. It shook me out of poetry in a weird way – not in a bad way, in a really interesting way. It shook me out of how writers live off imagery and cliché and as a dramatist I just started to really hate that because the drama is about this person and this person and what they're not saying.

JS: That process really worked for *On Blindness* but you couldn't write like that all the time. Image and cliché can be really important. I would also argue that BSL translation doesn't always translate, so when you are in the position to have the writer there in that collaborative process, it's extraordinary because you have the writer saying it means this not that it becomes part of everyone's language, if you sign or you don't sign, but the words of the play are crucial, they inform everything. Everything feeds into everything.

JM: Which moves us on to the way Graeae work, which is interdisciplinary, hybridized performance. Would you both talk about the play between 'texts'; the play between the human body and the written text, the spoken text, signed text, the danced text and so on?

JS: I think we cannot separate those things out. I think very early on there was a sensitive awareness and understanding of the dialogue that we needed to have. They weren't always going to be nice dialogues. Just thinking about a diversity of audiences and how can we enable them to 'get it', how can we create a commonality of experience? Just thinking about Julian's drawing of the set and then taking it away. How can we do that, evoke real equality amongst our audience so they can all experience it with the same passion. That dialogue was ripe to be informed by a human ethos of wanting to have a go. All of those layers, the sound, the movement, the audio-description and so on, they all informed every single conversation. It informed every line, every stage direction.

GC: There was a forced connection between disciplines in terms of where the material came from. For both Julian and Nick [Powell], for instance, the music often came from the words of the audio-description, so the rhythm of it was based on that so he was having to complement the text and time it with what Julian was doing on the video –

JS: and the timing of how Julian drew was in time with the music so that the deaf audience could get a sense of the rhythm of the music.

JM: So there was a sense of composition and choreography that was incredibly vibrant and tangible in process and performance.

GC: Yeah and because everyone was out of their comfort zone to some extent – Julian and Nick both embraced that, they were there in rehearsals as much as possible. It's also interesting seeing the results as well, how people were doing things very differently from usual, experiencing new ways of working. Also with Scott and Steven, it was really interesting for them in terms of their process; it was a real challenge to them. It was actually the best acting from them, their movement was fairly limited but they hated it.

JS: Scott and Steven hated doing it –

GC: They were really frustrated –

JS: I think it was more naturalistic than they wanted it to be. But I think the movement was just right, it was detailed, it was precise, it was maverick but the middle dance bit, we never found a way to audio-describe that. I have the answer. You know when you find yourself relaying the events of the last week or whatever and you replay snatched conversations you've had with people, and I was thinking that you could have overlaid that sequence with snatched conversation that we've heard so far in the play. That might of worked nicely. You live and learn and we went as far as we could, but there's a whole other performance, a whole other way of doing it always. Yet the central [Jenny makes a sound and facial gesture that suggests something inarticulable] of the process was perfect. The understanding, the not having to explain. I never ever felt like I was worthy, standing on my soapbox banging on about access.

JM: On the Graeae website you've referred to 'the aesthetics of access' and for me I think that's something about the [I repeat the sound and facial gesture] that you were referring to which is about being human and being sentient beings first and foremost. Is there anything in that approach that allows for something new to be uncovered in terms of ways of working and creating new styles of performance?

JS: I think the genuine desire to ensure that the work is accessible for all blind and visually impaired people and all deaf people at the same time is always central to everything. I'm in the luxurious position where, if I'm not doing a co-production then all my actors will be disabled. They come to work with me with different sensual and physical requirements. All of the elements from the design of the costumes, the aesthetics of the performance, it comes from them. That linked with the text. It's always a case of 'now what can we do here'. With *Bent*, every character was played by two people, one voice, one BSL. It's fabulous to be aware of all the layers of every word.

JM: And an incredibly rich experience for the audience.

JS: I'm in the best position because I have these fabulous actors who nobody else wants – someone may have a different voice pattern, how do we make sure people can hear that? Not just voice over them, but you also have to work so that the audience, especially a blind audience, can hear what they say. Even if you use traditional audio description you have to do it true to that character. It forces me way out of my comfort zone every time to think, okay what's the aesthetic of this play?

JM: Finding new and imaginative, wholly playful methods for work. You've both been talking about that very real sense of play that is at the heart of the work.

JS: Absolutely. I think one of the best things about Graeae is that you can't pigeonhole the company. Every single piece of work is different, by virtue of the performers. It's sometimes easier to get a following if you do the same sort of thing all the time. Our audiences are a real mismatch. And that's good too.

Details of the performances referred to can be found at: http://www.graeae.org/

2.9
Sara Giddens and Simon Jones of Bodies in Flight: The In-Betweens, Where Flesh Utters and Words Move – on Flesh, Text, Space and Technologies

Sara Giddens: Bodies in Flight is fascinated by and concerned with the encounter between flesh and text, particularly, but not exclusively, in live performance. Bodies that utter and words that move is a slight variation on our copy for the CD-Rom [*flesh and text*]. Coupled with that is our desire to energize the space between and in-between performers and spectators. Where spectators may be placed physically right up close to performers real and sweaty bodies. The work comes out of long-term collaboration between myself as a choreographer and Simon as a writer and equally importantly the various collaborations with a huge range of other artists.

Simon Jones: I would add to that, it's a kind of truism I guess, I was thinking of Tim Etchells talking about being interested to do things in the theatre that couldn't be done in other media. Our particular take on it is opening out these gaps between the senses. I think to conceive of hearing and seeing as two discrete channels of communication when obviously the normal tendency in theatre is to imagine, because it's in real-time/real-space, that the experience of the senses is continuous and what tends to happen in other media is that one particular sense is foregrounded; either seeing or hearing and the gaps between them are artificially created. It's interesting to think about those things as a wholly embodied experience for the audience. That's why the density of the material is something that we play with because in the real-time, the

Figure 2.6 Polly Frame in Bodies in Flight *Skinworks* (2002). Photo credit: Edward Dimsdale

dynamic time, of the event, we know people cannot process the amount of information given. In that sense it's a kind of deliberate overload.

Josephine Machon: This is central to (syn)aesthetics, that we are made aware of this fusion of the senses with certain types of performance work. You refer to 'the in-betweens' in terms of your collaborative approach, I'm interested in the way in which you pose that concept as a question; the way in which you put different languages together becomes an interrogation. For example you question 'the tongue and the camera'. What are the particular in-betweens that you have identified in your own practice and what exists within those spaces?

SJ: With that particular phrase, when we started, part of it was that there is such a thing as skill and expertise within practices, and that led us on to think about how practices each have their own language, each have their own discursive field, to use a Foucauldian term. We would want to push each discursive field to its limits rather than be interested in resolving a kind of overall meaning. We would pursue each of these different practices on their own terms. The in-betweens put us as makers into a strange no-where/no-place with respect to the material and to the audience-spectators. The everyday situation established is designed to offset that to a degree. In effect both of us as makers are as uncertain as the most open of audience-spectators to what the work actually does or says in the event. We really don't know it until we have felt or experienced it alongside the audience-spectators. In our attention, we push towards and hopefully *into* these in-betweens, between each other's competences and comfort zones, and that puts us into the no-place/no-where, where we hope the audience-spectators will also want to go. This uncertainty is exciting and we cannot begin to get the measure of it until the work is shared with an audience. So the making is profoundly about both sustaining a mood of being in that particular mix of in-betweens and also trying to shape it into an approachable event. The everyday helps us with that; but at no point do we occupy a position of knowing the event, until, that is, we have been through it a number of times with audiences. I guess this is a more radical version of how any set of collaborators come to know fully the work they have made. However, for us, the 'beyond-competences' is the zone where we are mostly at work. We began by rejecting machinery, what's called technology, because a lot of people were doing it and we felt that there was something being missed in the way in which that technology was being used to attend to surface; to surface imagery, bricolaging imagery, found imagery and we felt there was more in the actual working with bodies, with people devoid of machinery. But more recently, since 1998, we have used machinery and we have worked with sound artists since '93 who have used computer based sound.

JM: You've been using machinery quite extensively in your later work; particularly exploring virtual space and virtual 'relationships'.

SJ: But I think 'the tongue and the camera' was a backwards, retrospective recognition that actually the body itself is made up of a set of interlocking 'technologies'; that the tongue is also a technology. More recently we've been led to this idea of what we call 'second naturing'; the idea that there's a point at which the body learns how to use its own technologies and then in our contemporary world it dovetails into all these external technologies fairly effortlessly as well, depending on your generation and what have you. There's that dovetailing of natural and artificial technologies and that's what's led us into what might be thought of as explicitly multimedia work. We're beginning to think that this is all a spectrum of technologies from how you walk, how you speak, to how you might operate a digital camera or go online.

SG: The other thing I would add to that in relation to seeing things as technologies is about the excitement when something collides, like Simon's text and sound, so that when we work with Angel Tech, we went in with the intention of creating an opera for BAC [Battersea Arts Centre] but what was really exciting was working with that sonic material in relation to the text and the whole way that that soundscape allowed the text to hang, or build, or reduce. It was those *relationships* that became really exciting.

JM: So central to your work is those fusions, those active relationships within the hybridity.

SG: We became really excited by what could happen when you put these sonic elements in relationship to this rich text and that became the start of another piece of work. I also had that with video in particular, and photography, in relationship to what the still and moving image can do alongside that fleshy, live body. It's all those various interconnections, it's about colliding and colluding and it's as exciting when one enhances as when one undercuts or undermines. It's that foregrounding and backgrounding that is the stuff that excites us.

JM: These fusions bring me on to a feature that's central to your work, the idea of flesh that utters and text that moves; how the text and the body speak together. Expand on your particular areas: Simon, your approach to writing and the sensual and the visceral that exists in the verbal: Sara, if you would talk about your approach to choreography, your very particular style of 'micro-choreography', the tiniest nuances of the body and how it speaks.

SJ: Two things exist in the writing. One is a kind of forensic investigation, which really began with *Do the Right Thing* asking where are we and what are we doing and what are we looking at and what are we able to say about it, trying to investigate the situation as thoroughly as possible. That situation, certainly with *Do The Wild Thing* was rooted in the theatrical, that 'scene' of desire and scopic investment that theatre is. Theatre is this thing where people look and listen to other people. So I will do the bit that people listen to and Sara will do the bit that they look at. That forensic nature of the writing completely changed the nature of the earlier texts which in the CD-Rom we call 'satiric' in the sense that the earlier texts were comments on being alive in the UK in the early 90s. The second element that's developed more recently is an interest in the sonority of words. Poetics is the wrong thing, it's more a kind of going in and around and through little circuits of words which almost start to detach themselves, start to have a kind of sonority to them that is separate from their [semantic] sense. It's that combination of intellectual sense and sonority which is the form of poetry I'm doing I guess. I've become very interested in the palette of the vocabulary getting narrower and narrower.

JM: It's interesting that you use the terms 'palette' and 'circuit' in relation to writing, like you're playing between being the artist and the computer programmer.

SJ: I think of circuit in the sense that it goes through time. By circuit I mean that, now, one of the things that I do when I'm writing is that I try to forget that it's got to be spoken, that it's got to be performed. I'm deliberately setting a challenge for the performer because it's not dramatic in the sense that I'm not writing to particular strengths in performers that I know or trying to create a character. The first part of the writing process for me is a bit like clearing my head of everything and trying to write into a certain language, a certain text if you like; I try and find the text. It's a bit like stalking it. I don't have any ideas for what I'm going to write or going to say, I don't have any plan of writing, stuff just accumulates over many months and it's like it coagulates around certain themes and repetitions and circuits. Then what I do with the performer in rehearsal is unpack and find a way through that material that the performer can then speak. Obviously they have to be able to speak it but it doesn't begin as stuff that is speakable in that sense. It's written but it refuses to make that jump; it doesn't slip easily into the speakable. The rehearsal process is about making it palatable, literally in terms of the palate of the voice but also palatable for someone to listen to.

JM: So does it become another artefact that is something that you might play with in the rehearsal process, to discover what might exist within it?

SJ: Yes. A standard play script is written with the total intention of it being said and if the writer's any good it will only require a little bit of tweaking and there's a set of conventions to follow. One of the interesting things about *Model Love* has been working with two performers who have never handled the text before and one performer who's very familiar with the way that I write and it's very interesting that those two performers have had difficulty with it in trying to apply their notions of what might be poetic, dramatic text. They've found it quite difficult to find a way into it and a palatability for the writing. I think it's only by running up against the unpalatable that it becomes interesting and that's to do with tearing apart, or troubling, the difference between sonority and sense. It does seem to be going more minimalist and less baroque.

JM: So it's becoming more like Sara's choreography in that respect. It always used to feel like yours was the text that was doing the dancing and Sara's was the text that was being quite still.

SG: I was just thinking in relation to that about how 'bespoke' the physical language that we use is. How I choreograph is about detail and working with the performers' own bodies and very small nuances or minutiae within their own physical vocabulary; really focusing in, honing in on those moments. Often actually 'stilling' those moments, holding those moments in order to then be able to build up a choreography, a vocabulary from that. It's about finding the moment and then being able to recreate that moment and quite often repeating it.

JM: And do those moments come about as a result of this collusion and fusion of ideas and disciplines in the rehearsal space or is it specifically through something that's found in a particular body and how it might be responding at a given time to a particular moment in the text?

SG: It's very particular. It's a particular body at a particular moment. We might set up improvisations that are seemingly loose but have quite tight structures around them. All the time it's about looking for those moments that are then stilled. Working out why I am drawn to those moments; what it is in those moments that is drawing me in terms of my looking and in terms of my *feeling* as well. Then, how are the performers approaching those moments? It's a mutual development in that sense. I might suggest that they jam within a certain key moment and we find that material together. There's a dialogue going on which is about them moving and me watching and then me physically suggesting ways that they might want to approach it and occasionally teaching really small parts of movement. It's bespoke, which means that it changes if another performer comes in, those movements are changed.

JM: When you talk with your audiences do you find that individuals focus on a particular area? Do they hone in on a particular language that you're exploring or do they talk about the total aesthetic?

SG: Quite often the response is about where they come from and what they find it easiest to talk about. What's interesting is often it's those people who aren't familiar with a specific medium who are most able to talk about the whole experience. Often I will be with people who are movers, dancers and choreographers and they will talk to me about that aspect of it and Simon will be with people who are more used to texts and plays and they will talk to him about that. Whereas I find that those people who haven't got either background can talk about how they felt, how they experienced the work as a whole.

JM: Following on from that, do they talk about where in the mix they believe the *ideas* exist? Arguably there are those more traditional theatregoers who believe the ideas exist in what is spoken. Do your audiences talk about the ideas within, around and behind the work, within the fusion?

SJ: I think, because these channels of communication, or layers or slabs or dimensions, that's probably a better word, the dimensions of material that we're using don't resolve, very often people think they're missing something. That's because they're so used, even within so called avant-garde work, to being able to resolve, even if the resolution is 'oh this is multimedia' or 'this is an open ended text'. Our performances don't feel open ended they feel like they mean something but then it's very difficult for people to *say* what they do mean. There's this strange kind of halfway position that some people find quite irritating and frustrating for different reasons, either because they're used to being able to describe the event, come up with a tag-line for what they've just seen or because they're used to a particular kind of performance and this doesn't quite fit into it. I think that's why we started to work with what we call 'the situation' more recently where we've tried to find the induction in the classical sense; a way of leading people into the work which is about everydayness, trying to take them from their everyday world into the Bodies in Flight world. It's like a portal that allows them through.

SG: The hook is often 'the everyday'. We talk about that, not just taking them through the portal but the hook itself that allows them access, that being something that we all experience like loss or love or longing. That's about a 'consensuality'.

SJ: I think that's what lies behind Sara's statement about it depends on how people come at it. They'll find that they will have an openness to it because they immediately associate with the 'stories' from either discipline. *Who By Fire* begins with a half hour conversation about weird deaths and after a few minutes almost everyone 'gets it'. It is a classic theatrical seduction in some ways; the performers seduce their audience. The stories allow people to open out part of their sensitivity as spectators and to be open to the material as it opens out yet doesn't then resolve. It doesn't always work in that sense because we resolutely do not resolve the material. We open the door, it opens out and it's true that a lot of people, even those inclined towards experimental work, don't like that.

JM: There's something interesting in what you've been saying about how the audience become aware of *how* they are engaging in the moment as much as following the event. It could be on a sensual level or an intellectual level, or combined.

SJ: When people are prepared to say what they got out of it, they often have got quite a lot, but when people try to second guess what might be going on as an uber-idea, a super-objective, then they get frustrated and that's because we deliberately don't resolve those things. At no point do I say that this line has got to go here because it makes complete sense with this move or this image. There are no hard connections between the different elements. It's more like resonances. The word we've been using a lot lately is 'mood' because it implies connections between things but doesn't say they translate. We've always quite rigorously said you can't translate a move into a line or a line into an image or an image into a bit of sound; none of these things are translatable or exchangeable in a commodity sense. You can't line them up into neat packages of this word equals this move equals this image, it doesn't work like that. We've felt more that the shows create a mood and the situation of the show. *Model Love* is about photographs and the taking of a photograph that might be the moment where someone falls in love, might be the love at first sight photo. It's a situation to invite people to think about what do they really get out of photographs. How many people really know what they get out of photographs? I know that I don't know, the more I think about them the bizzarer, as objects, they become and yet they're everywhere. With *Who By Fire* it was why do people at three in the morning, with strangers, end up talking about death? It's about these everyday things that when you choose to look at them you suddenly think, how strange is that. How strange is a photograph?

SG: It's the idea of creating space around that. Look at this or hear this and then create the space around that or out of that for people to connect to it. That was really clear in *Who By Fire*. We had created a space in *Who By Fire* where audience members were able to come in and out of focus to it. They weren't frustrated or annoyed about being able to float in and out of the sound or the text or the movement. With the latest piece, they are much more able to pull focus as *that* is the central idea in *Model Love*, that moment of focus on a particular image; what happens before that moment of focus, what happens after that moment of focus. What we did for the audience was allow them to have that space. Quite often people feel like they mustn't allow themselves to drift in and out of it.

JM: You've just talked about drifting in and out of spaces, journeys and portals and being taken into the Bodies in Flight world which is interesting as something that exists in a lot of writing around your work, whether by you or other people, is this sense of a journey that is taken through it, even if your audience is sitting end-on to the performance. Also there's a shapeshifting quality to your work, which isn't just about each piece, being a work-in-progress but about decisions being taken to remould and reshape at each performance. Each becomes both something other and yet still 'the same' so there's a feel of a journey in this shifting form, a playfulness at its core. I wonder if this is something you're moving further into; *Model Love* is a durational installation for example.

SJ: Durational in the sense that it's an ongoing installation that lasts over a period of time and the audience is invited to come in and out, obviously they could stay for the whole four hours if they wanted to. We did do that once before with *Constants*.

JM: Which was about time and aging so the content demanded that form.

SJ: Yes, and I think the motivation with this piece is twofold. Firstly that we wanted to make something that was more focused on the photographs, to give people the chance to experience them in the way that they might normally experience still images, they're allowed to look at them for as long as they want to look at them and then move on to the next one, that gallery experience. Secondly, with *Skinworks* we by chance discovered that however we (re)combined the elements of *Skinworks*, something of the essence of *Skinworks* always remained. We wanted to try that again with *Model Love*.

JM: So that shapeshifting quality plays with traces and ghostings that exist in the work. In *Model Love* you're exploring the idea of capturing a moment in time and yet by doing that you're constantly engaging with that moment in time, making it a constant presence that's always in the process of becoming something other at any moment.

SJ: As Sara says you're always *attending* to it. In that sense the material, in my case the words, becomes the means. I may get pernickety about the words at a certain point but before that point they are very flexible. They will change gender or person, sometimes I'll write in the first person and that will be shifted to third person so the actual grammar of the material changes as we work on it in rehearsal. Strangely the actual material is a means to getting to that mood.

JM: Which is how Sara talked about the grammar of the body, always shifting and changing dependent on the bodies involved. Once choreographed it is also always in the process of shifting because the body is always doing its own thing.

SG: Absolutely. I was just thinking about the rhythm of an installation like this in relation to how we have worked on shows. We always have an absolute excess of material in every channel so often for performance we're paring down. With this as an installation the paring is extended over that period of time. I find that really interesting. It feels like a really good moment to let the material breathe for a while before, I suspect, part of it will be reigned in again for something else. This is what Simon's talking about in terms of the absolute essence of the work that we find or that rises up; in the same way that it is in any of our theatre pieces. Those same resonances are still very present but they're worked at or developed in a different way; or allowed a different space/time quality or context.

JM: So the form morphs but that thing that's at the heart of the work remains constant; which is something to do with a 'sensual conceptuality', sensual ideas remain at the core.

SJ: That's one of the depths of *Flesh & Text*. On one hand it's a glib title for everything we do but it is that notion of the embodied and the discursive, something that's felt and something that's thought. Again, it's all these gaps and in-betweens that we cover over in our everydayness. The privilege of performance is that you can break those things up for a while and just let people experience the gaps, attend to the in-betweens, which ordinarily you can't do because they're not very helpful to everyday life.

SG: The word that we use is 'dwell'; to dwell in those particular intersections or to dwell in one particular mode. I find that a really useful way of articulating where we want to be or where we might want an audience to be. We talk about that a lot when working with the performers; just allow yourself to dwell in that space, in that material for a while. What we do takes a lot of time to do it. We set up long improvisations and the work comes out of that longevity.

JM: I want to return to the idea of the in-betweens in relation to the thinking and the doing. *Flesh & Text* is a document that illustrates the fusion between these areas, the in-betweens, illustration that you invite critical reflection about your work. Would you finish by talking about the value that you see that occurs in that space? Within that, how key is the corporeal experience of the work to the cerebral reflection on it?

SJ: Sara's sojourn in academia was just part of her career whereas I have never been out of it. To talk personally, it gives me a structure, security, a space to go off and think things. Being paid to be an intellectual means it gives me the range to do those things I want to do – it's very luxurious. I've never taken the view that thinking about something potentially destroys it or damages it. That was the traditional wariness of artists working with the academy, that if they submitted their work to the forensic gaze of the critic then somehow their work would be destroyed. I guess I automatically have to work things out and what interests me is that I *can't* ever finally work it out. However much I look at the same questions, rather like a fractal image, the depth of them is infinite.

SG: I think the thing is where you place reflection time in the work. Often the opportunities to write or talk are created separately and that's where the reflection takes place. When we're making work we spend an awful lot of time outside the rehearsal studio talking and thinking and reflecting. As long as the work is rooted in everydayness in some sense then that allows one to be quite philosophical about it. It comes out of this combination of intellectual and emotional and felt and thought 'stuff' and that carries over from one thing to the next.

Details of the performances referred to, including the CD-Rom *flesh & text*, can be found at: http://www.bodiesinflight.co.uk/

2.10
Leslie Hill and Helen Paris of Curious: Embodied Intimacies – On (the) Scent, Memory and the Visceral-Virtual

Helen Paris: We talk about the work of Curious as focused on contact and communication and that can sound quite bold but actually those themes and that ethos really underpins the work. Of course we work in the digital as well but if you are committed to working in the live then I think that there are things that drive you in an almost evangelical way. That idea of what contact can you make, what is palpable and possible in the live moment with your audience? That is really strong for us.

Leslie Hill: I guess the relationship between the visceral and the virtual for us is partly because we formed our company in 1996 so we were looking at the virtual in terms of the internet and virtual reality and those technologies that were new at the time. What is still true to the work that we do now, in that a lot of the work that we make is about quite intimate moments of contact in the live. We're always looking at ways that you might be able to use new technologies to deliver an intimacy to an audience that is beyond the live moment. I guess a lot of the projects are balancing acts between trying to create something that is intimate and of the moment and looking at strategies for using new technologies to allow greater access to the work. *On the Scent*, for example, is for audiences of four people at a time. Even though we've done the show in 16 countries there will always be a limited number of people who will actually have seen it live. The visceral-virtual also describes how, even though there's no substitute for the live presence and the one-to-one moment, there's a lot of interesting things you can do with broadcast technology or digital and online technology. Also publication is a one-to-one way of having

Figure 2.7 Curious *Lost & Found* (2005). L-R: Helen Paris, Leslie Hill, Lois Weaver. Image © Curious/Arts Admin. Photo credit: Hugo Glendinning

contact with language. Sometimes publication is a nice way of opening the work out to other audiences.

HP: I think that word 'virtual' is interesting because on first glance it implies distance but actually it's also about what's palpable, what's 'almost there'. So in terms of the virtual visceral, I like that double meaning of the word, it's what's absent and present at the same time.

Josephine Machon: Would you talk more about the 'intimacies' that exist in your work, in particular the one-to-ones you've mentioned? What are the aesthetic or conceptual ideas that produce those moments? What are your intentions with that intimacy?

HP: *Vena Amoris* was the smallest audience we've ever had, just one person at a time, one performer and one audience member. What's really interesting there is, it's the thing that Peggy Phelan says about performing becoming itself through disappearance, with *Vena Amoris*, because the audience is led on a journey through a building and the performer is not with them physically, visible, although the audience member always feels watched, which is really interesting as there was only one moment of actual meeting and still that's separated through a two-way mirror. I think that it's that sense of it being more intimate, more palpable, more visceral. One of the most extraordinary things about that piece was the feedback from the audience who felt they'd had a hugely intimate, hugely physical, hugely visceral experience with the performer even though it is only a phone conversation. There were some really interesting thoughts about what that intimacy is on the phone exchange. For me, the technology of the phone is so beautiful in that it's quite old-fashioned and in this piece that idea of distance allowing more intimacy. It's a really interesting connection between the intimacy of performance and the intimacy of the electronic, wireless, digital media and how it can engender even more of a meeting.

JM: How does that intimacy affect you as the performer in terms of getting back to the basics of performance; you interacting with another body, others' bodies; that simple, direct, straightforward manifestation of performance?

HP: For me, being the performer in that piece, it was hugely compelling, I felt addicted to it. I felt that I only ever wanted to work with one-person audiences again. The next piece that we made after *Vena Amoris* was a piece called *Deserter*, which was a piece commissioned by The Project in Dublin. In that piece we tried to find another way of creating that kind of intimacy in a black box, 200 seat venue. We did it by having an audience of 40, so there was a still a limit on it, who entered the theatre space one by one and had this moment in a confessional booth –

LH: it looked exactly like the theatre door. They were told they had to enter one by one but as soon as they went through the door they were in a sauna and it had the sand that's used for glass-making on the floor and it was really hot. But it was also a confessional booth. The grate slid open and Helen confesses to them and it slides back and then I open the door on the other side and they come in to the theatre.

HP: So it takes 40 minutes for the piece to start because of this one-to-one moment that they have. In a sense that moment of the grill reflects back to that moment in *Vena Amoris* where there's the meeting of the audience member and the performer face to face through the two-way mirror. That part of the piece absolutely came out of that desire as a performer to still want that exchange and to see what intimacy you could have in a large audience.

LH: I had the hard job of being on the other side of the door because the one-to-one moment was great but then how do you keep the rest of the audience entertained before the show starts. I was doing some text but it turned out that because this one-to-one moment was strong for them, they were just fascinated to watch everyone come through the door, especially people who'd come together.

JM: I've noticed that with other work the delight that is taken by other audience members in the reaction and experience of their counterparts; in the palpability of that experience, that sense of discovery that they're tapping into.

LH: That relationship between the performer and the audience, for me when we do the small scale pieces, it's such a relationship of trust with the audience. When we do theatre pieces you can have a good audience or a quiet audience and you know the feeling of a great performance and an okay performance through what you feel off the audience but in a one-to-one it's such a delicate balance of trust because they can topple the performance at any moment by how they react, yet it's also really full-on for them. One woman said to me I felt really on it because I was a quarter of the audience so there's this really interesting kind of –

JM: responsibility?

LH: yeah responsibility. It feels like undertaking a joint venture, more than a theatre performance.

HP: Some of those pieces literally manifest themselves as journeys; *Vena Amoris* was a journey around a house, or *Lost & Found* where we have a slightly bigger audience of 12 where the journey is on a boat or on a bus but the actual journey that an audience and a performer do go on when it is that intimate, very up front, is really something quite particular. There's something very interesting as well in that there's an afterlife to each piece. We really know who our audiences are. Partly because, like with *On the Scent* at the end of it the camera that Lois Weaver is filming herself with at the beginning is turned on the audience member to give their own smell-memory, so we have this huge archive of 3000 smell-memories from all round the world. It means that, for Leslie particularly because she's spent all this time editing this footage to make DVD output with them, there's a real sense of 'knowing' when we see people again, like when you recognize somebody famous and you think, 'who is that?', it's the same thing with our audience members.

JM: Key to your work and part of this incredibly intimate relationship with your audience is the idea of 'body knowledge' and I know this is something you're working on specifically with *Autobiology*. Would you talk in greater detail about that in relation to the embodied aesthetic and the consequent embodied experience of the audience?

HP: As well as contact and communication the overarching theme or concept or belief in body memory has suffused a lot of our work, sometimes more obliquely, sometimes more specifically. Body memory is a way that I would generate text for myself. I'm really interested in literally finding what text comes out of movement that leads to sound and then what text comes out of sound; finding different ways of generating text via a store of memory in the body. I'm really interested in cellular memory and all the Eastern traditions of how we store histories and legacies, tension, stress and emotion in our body and there's various ways, like yogic ways, of releasing or tapping into those. Creatively, I think it's a really interesting way of making work. Our work with smell was a whole different trajectory that came out of working with body memory because of the huge trigger that smell has in terms of memory and emotion. With *Autobiology* we're literally going in to the gut and looking at what that is when we talk about 'gut feelings' or when we talk about making work 'from the heart' and how can we get in there and find out what that holds.

JM: What does it add to your practice to examine the actual physiological, anatomical roots to those metaphors?

HP: It adds something on a totally different level of intimacy, of how we know our own bodies or don't know our own bodies. That physical relationship that you have with yourself, with your own body externally and also that complete lack of knowledge we have of ourselves internally, combined with that instinctive place where you 'know' yourself. I'm interested in finding what that is. In some ways the work with smell- memory was also doing that, because as much as the memory feels intangible, it's also intensely visceral.

LH: Every project has different points of contact and communication; sometimes it's through one-to-ones, sometimes it's through us making a theatre piece and then we make a film as a flipside and further contact and communication happens through the filmmaking. In *Autobiology*, a lot of the really direct contact and communication is going to come through a series of workshops. We're going to take methods that we've used for making work for the last ten or twelve years and put them together in a two-week long workshop as part of this project, specifically focusing on body memory and research. We've both been doing a lot of research, reading scientific books and so forth and now we're working with a neurogastroenterologist. We really wanted to look at this idea of gut feelings, making decisions from the gut, *knowing* from the gut, the shared human experience of that term. We want to get to the root of that term, what really is happening there. In the same way that with *On The Scent*, we did a lot of research with scientists there around the way in which smell leapfrogs the hippocampus and it's not processed in the same way [as other senses] and that's why it gives you that really direct feeling. We're interested in the universality of 'gut feelings' but also in the fact that a lot of recent scientific research is overturning the Cartesian idea that the head is 'cut off' from the body. Science is proving that the body *is* the brain, every part of your body is the brain and how you think and feel is much more bodily than the Western world has been thinking for a long time. The gut is a really great example of that; it's the only part of the body that has its own second brain. People on life support, for example, don't have to have their gut monitored because it can work totally independently of the brain. It's very rudimentary, like a worm's brain. It's also lined with peptides, which are like little bits of brain floating around your body so when you feel something really strongly in the gut, it's because it has more peptides than anywhere else. We're really curious about all these things and wanting to use biofeedback is as a result of wanting to turn the body inside out. Our fantasy for ourselves in performance and these workshop environments is to engage with how far could we interrogate this and how much could we show to people? We've been taking part in experiments in a lab as a way of getting to know the scientists better but it also allows us to see the various kit that they use in action. There's one piece of kit called a neuroscope that we thought we would most want to use but then I did an experiment where I had to have a tube shoved up my nose and looking at the neuroscope feedback, audio-visually, I was very disappointed because I thought, that's not going anywhere in an arts context! Some of the incredible imagery that's coming out of MRI scanning, you think, give me something more like that! So the research is partly wanting that information but also partly wanting to –

JM: to find an aesthetic?

LH: to find an interesting way of presenting it. To an artist, that little graphic isn't going to do the same thing that it is for a scientist. So we're in the middle of trying to figure out biofeedback interfaces. There's a lot of performance and dance that already uses it, so we'll talk to other people about what they've done and what worked or didn't work. It's basically a desire to do science-like experiments with ourselves and with other performers who would take the workshops and then, ultimately, in a performance situation where people from the audience could try it. Another reason why we're working with this lab is because they use autobiographical writing as a research methodology. They have people write autobiographical things and then they test their gut reactions when someone reads back their own stories.

JM: So there's a beautiful symmetry in the collaboration.

HP: Totally. We wouldn't call our work autobiographical but that notion of generating work from one's own personal experience or from the body's autobiography; that makes the 'autobiology' of autobiography. In part it's a conceit and in part it just might not be a conceit.

JM: So it's as much about finding a form as it is about exploring content. This is illustrative of the fact that at the heart of your work lies sensual concepts; the thinking and practice is an interrogation of the sensual and visceral. Would you talk further about that and about the tangible, sensual metaphors so prominent in each piece, where form and content fuse and articulate each other?

HP: I think you're right, that is there explicitly or implicitly in all the work that we do. *The Day Don Came With The Fish* was very much to do with form meeting content. We were using technology in terms of mortality and immortality. The content of the piece was about a friend's death and about a friend living with HIV, the idea of a moment in your life changing everything, when the infection enters the body. Using old and new technologies; Super-8 film, the rupture of an Edith Piaf song on a 78 rpm record, analogue and digital code to play on the code of the virus, but also to do with time and rewinding and trying to go back to the moment before everything changed. Also this idea of the immortality of digital media and the mortality of Super-8 film in the performance, which burnt on the projector so it always has the burn in and on it, and similarly the scratch in the record. The flip side of that is that in digital code there's always that speechless moment but in analogue there's never the rupture. So those twists and turns and how *all* of that is played against the body and on the body, projected on the body, about the body, so that you're looking at the mortalities and immortalities of digital and older technologies in the context of our own mortality and immortality. It's like the technology becomes a permeable membrane between the performer's body and the text; it's there in the thinking, it's there in terms of form and content.

JM: Another key feature in your work is hybridity; its multimedia form. Would you talk further about the multilayered metaphors, juxtapositions that exist in this approach and why you make those aesthetic decisions to work in that way?

HP: The reason we called ourselves Curious is because we're always interested in the questions and we go wherever they lead us. All of the projects have a lot of research behind them so I think we find the form that holds, or works against, that content. So it could be that this will be a one-to-one performance or this will be a piece on the web or this will be an installation because it's always being driven by the content and how we want the form to *inform* that, or create friction within that.

LH: I think it's harder but ultimately more interesting for us. I think in some ways it's harder to be content led and to see where that takes you. It's exciting because of the flexibility of the outcome of different projects but it's also difficult as practitioners to keep up with all these different fields and to be 'expert', although then you bring in collaborators to work on different projects. In the early days when 'new technologies' were newer, we did do all our own editing and web-making and tried to be masters of all these trades. Now they're so specialized and so much further advanced that, as much as I do all the film editing, we tend to get collaborators so as not be limited by our own skills. Sometimes I think it sounds tempting to just focus on one form but it wouldn't be true to us.

HP: It feels totally instinctive. We met by seeing each other's solo work and our solo practice was really quite different from each other; *how* we made work as much as the work that we made, and that was part of the attraction. The first piece that we collaborated on and made as Curious was a road trip that was updated on the early web, *I Never go Anywhere I Can't Drive Myself*. It's really interesting to see the work we make together, what the catalyst is. We always have a continuum in the live but we do shapeshift. When we made our first film we were making a piece about longing and it just felt that the right medium to talk about longing was 35-millimetre film. Every project we do often has three component parts to it; there'll be a lens-based part, a live part and an interaction with the audience. That's been the continuum for the last few years, the audience becoming, in some way, participant in the work.

JM: So as well as producing a shapeshifting form, the playfulness allows for a sense of discovery within the work, for you and your audience, which leads you on to a journey into the next piece. In relation to *Gut Feelings* the film that will result from the *Autobiology* project and *Essences of London* [the film document that partnered *On The Scent*], would you talk further around your approach to the sensualities of screen; how you manipulate film in a sensual way, how your ethos works within that format?

HP: It has a lot to do with what you were saying about the visceral-verbal quality of text. One of things that I love most about working with smell and specifically on the *Essences of London* project where we went out and interviewed people, often in very olfactory trades, is how when people talk about smell they become poets. It was extraordinary to me some of the things that people said. We went to all sorts of places in terms of location in London and therefore in terms of class, culture and ethnicity. We had a real range and yet every time that way that people talked about smell became poetic. I absolutely believe that's because the memory is coming not from the head but from the body because it's a physiological memory. You can often see people being transported as they start to recount the story, the memory; you see them going back to that place, going back to their grandmother's kitchen sink or their trip to the hop farm. We had garbage men who were teary eyed –

LH: big, tattooed bin-men talking about their childhood trips to a hop farm in Kent –

HP: and you might have an expectation for that with older people where memory is incredibly pertinent but we also went into teenage homeless shelters. We interviewed a lot of big, tough guys at the market and with all of them there was this moment of time standing still and them going on a journey into the past and (re)living it in that moment. In terms of performance and capturing that on film that's really rich. You don't have the actual liveness but you do in terms of the text; the recall generates a very particular kind of textual document. One of the things that really excites me about *Gut Feelings* is that we're working with Andrew Kotting. We've wanted to work with him for a long time because his films are extraordinarily visceral. *This Filthy Earth*, it's dirty in terms of its content but the way he employs film is so visceral, you feel dirty after watching it.

LH: How we've worked with sensuality and film before is more to do with the responses that we elicit and capture than it is to do with us using film in a sensual, Peter Greenaway way. We haven't yet used it in a visually super-sensual way but that's what we'd like to do with Andrew, bringing him in as a director, visual artist and filmmaker.

HP: Although I would say with *The Day Don Came With The Fish*, the idea of projecting onto the rice paper and eating that, where the filmic image becomes consumed, and projecting onto the body, there's definitely been that idea of working with film in that way.

LH: It is sensual but it's made so by what's happening in the live performance.

JM: I'm really interested in what you've just said about transportation through the senses and the idea of time standing still and (re)living memory. In one and the same moment that experience fuses the research, the performance and the audience reaction. It garners an (im)mediate appreciation of that performed moment. It connects with what Lynn Gardner said about *On The Scent* 'the ordinariness of the domestic setting is transcended and transformed into something quite extraordinary'. It allows the audience to experience the ineffable where that *feeling*, that heightened knowledge, becomes tangible, so transformation through metaphor as well as through actual visceral experience.

LH: With *On The Scent*, when we started on the project we knew we would research in India with the olfactory scientists and we knew we would have an output of some kind. Initially we thought it might be a museum installation where people could go through and experience different smells and stories. We weren't set on it being a performance originally because smell is really hard to broadcast, contain and contextualize. Finally we ended up thinking let's not use a gallery because it's too sterile in terms of smell and at the same time we were doing the interviews with people and so many of the smells talked about were domestic, so we thought let's use a house because that gives a framework that contains the smells and makes sense of them in the different rooms in the house; the cooking smells of the kitchen, the medical smells of the bathroom cabinet. Rather than an installation in a house, we realized that what was fascinating about the project was people telling their stories and watching the impact it had on them remembering. So we decided that the best way to present it was through performance that included an open-ended element where the audience could participate. Three performers, each in a room, using a metaphor of a personal experience, with the arc of the personal stories told via smells which allowed the audience to experience what we had found fascinating in others' stories; us being transported elsewhere through these memories but in a way that was accessible to them and everybody, through seeing it, it triggered something which made them always eager to share their own stories –

HP: absolutely, they needed to do it. That was another fascinating thing with that project, tapping into a core humanity in terms of smell crossing geographical borders. That 'border-crossing' sense of it suggested there were so many things that unified us as humans. We might be talking to someone in China or in Brazil, a really different cultural mix but that memory of the grandmother, there were memories that were key and overarching wherever we were, as well as the things that separated us, often to do with environment, nature, political eras. In the performance, obviously people are responding to some of the smells, like the perfume Lois used or the burning smells in Leslie's kitchen, or the Dettol smell in the bathroom. Also it's the text that they're responding to, text is generating the memory, so it's the environment but also the stories in that environment. In the beginning, I knew how profound smell was for me, a huge creative tool and emotional trigger, but I didn't know how it would be for an audience. I believed that it could have that power to transport but how to do that without them feeling manipulated, for it to be authentic. People respond to it authentically because of the bypassing of any censorship in the performance that we had made. The intimacy that we offered, that moment of contact and communication, became hugely profound and that's the essence of life really; to make contact, to exchange something in that moment.

JM: In terms of that authenticity, how does that connect with the visceral-virtual, because there's also an implication in 'the virtual' that it's removed somewhat from the authentic, how did you bring those two together in this piece?

HP: I think that it was really up front and visceral. Everything that we did, like setting it in someone's home as opposed to a gallery, we did everything we could to make it of the body, from the body and for the body.

LH: The virtual part of the *On The Scent* project was the DVD. We're trying to open out access, although there is always that part of you that wants everyone to see the live performance.

JM: It offers a different experience of tapping into those moments but what that says to me is that there's something about the live moment that you cannot replicate.

HP: Totally. We made lots of connections between live performance being like smell, it's 'unpindownable', ephemeral, you have to be in that moment and then it's gone.

LH: With the DVD I was really excited about the tracks with the Londoners detailing what London smelt like for them and giving smell-memories of their own and I was excited about the audience tracks from the performance, the five different houses in London, so the DVD project really worked for me on those levels as direct storytelling. We argued around putting a video of the live performance on there but the audience interviews and the out and about in London are satisfying because that's the format that they were conceived as whereas the live performance wasn't, so it's frustrating to view in that format. Yet we wanted to give a sense of the project as a whole –

HP: of what triggered all those stories.

JM: It's interesting that you refer to body memory as a springboard for your projects. For me corporeal memory is vital to work that functions on a visceral and/or transformative level so that as an audience member, when you recall it or make sense of the work intellectually, you cannot help but draw on that corporeal memory. The work is demanding that mind/body connection that you referred to and I do think it's possible to get those moments from film and for film to capture a quality of the live performance in a unique way. In terms of embodied knowledge you both also work in academic environments and we're all aware of the tension that can exist between practice and thinking. What for you is the significance of thinking, of theory, analysis, to practice? What demands would you have of analysis in relation to your own practice?

HP: Prior to being involved in any academic institution, that idea of what those questions, or theories or philosophies are, just approaching them as questions, has always been really significant in the whole body of our work. Like Leslie was saying, we have long periods of research. One of the things that got me into writing about my own practice was that feeling of the artist as struck dumb; they make the work but it's up to someone else to interpret it. And it is, it's the audience's responsibility as much as any formal analysis. However I'm really interested in how the artist describes their own questions, theories, philosophy. They are a really true and organic part of the process for any work that I make, notwithstanding me being involved in teaching in academia. There are other artists you've mentioned that will be in this book, like Akram Khan where the thinking that goes on in that work is really profound, really beautiful and you can see it in the work. In the work that we do and our contemporaries, like Marisa Carnesky, that thoughtfulness and those questions are really significant.

LH: For us, hybridity is our thing, so it makes sense that we have both thinking and practice combined. Sometimes I feel quite evangelistic about wanting artists to be more self-actualized because we work with a lot of emerging artists who can be very down on theory, or see it as something irrelevant to them and also down on doing funding applications which is another area where you have to articulate how you define and where you place your work. We really encourage those artists to take it on and own it as another facet to what they're doing. It's important to get in there and grapple with it and represent yourself and be clear. It's demanding, it's not easy to be a good theorist, or to articulate your work; it's no easier to do any of that than it is to make a good piece of work. They're all complicated and complex exercises.

JM: I love that idea of it being part of the hybrid, one of the many disciplines fused within the form.

HP: Also, I wouldn't pressure any artist to feel that they had to explain or justify their work but I wouldn't want an artist to not think that they are creating or working in contemporary theory and practice. They're part of that.

Details of the performances referred to can be found at: http://www.placelessness.com/ See also Paris, 2006.

Notes

Introduction: *Re*defining visceral performance

1. I use the term 'visceral' to denote those perceptual experiences that affect a very particular type of response where the innermost, often inexpressible, emotionally sentient feelings a human is capable of are actuated. The term also describes that which, simultaneously or in isolation to the emotions, affects an upheaval, or disturbance, of the physiological body itself, so literally a response through the human viscera. Following this, by experiential I intend those encounters or occurrences that we engage in directly which affect us on this level, intellectually and corporeally. The experiential in performance, following Gilles Deleuze and Félix Guattari, summarized by Susan Broadhurst, accentuates experience as 'unmediated and immediate' with its own philosophical powers. Rather than experience being mediated by ideas, 'ideas are extended by experience' (see Deleuze and Guattari, 1999: 47; Broadhurst, 2007: 38).
2. Throughout this book where I employ terms such as sensate, sensation, sensual or sensuous to describe works of art and specifically performance I do so to highlight how the work itself makes direct contact with the human senses. This in the same vein as Deleuze' use of the term 'sensation' in appreciation of art; transmitted directly it 'acts immediately upon the nervous system, which is of the flesh ... reaching the unity of the sensing and the sensed ... sensation is in the body and not in the air' (Deleuze, 2004: 34–5).
3. As Broadhurst has noted, there is 'a noticeable lacuna between such practices and current critical theory' (1999a: 1). Geraldine Harris (1999) also highlights the divide between theory and practice as experienced in performance terms, where the artistic work can be seen to be appropriated by the discourse, and reduced by over intellectualization. Arguably dance theory has gone some way to broach this lacuna such as is demonstrated in the writings of Horton Fraleigh, or in Roger Copeland's demand for analysis that presents a 'sensuous portrait of ideas' (see Horton Fraleigh, 1995 and Copeland, 1998). I agree with Harris that the ideal relationship between theory and practice 'is one of equal exchange if not interchangeability' where the 'perceived gap between theory and practice' can be seen as 'a potentially productive space' (Harris, 1999: 1–2).
4. For early published examples see Machon, 2001a and 2001b.
5. 'Shapeshift' here is taken from 'shape-shifter', naming Churchill's mythical underworld creature, the Skriker, which has the ability to morph and change form as desired (Churchill, 1994a: 1).
6. 'Writerly' follows the post-structuralist theories of, in particular, Roland Barthes, Hélène Cixous, Jacques Derrida, Luce Irigaray and Julia Kristeva. Raman Selden, Peter Widdowson and Peter Brooker define writerly texts as those that encourage the receiver 'to *produce*', and play with, 'meanings' rather than simply consuming a specific 'fixed' meaning (1997: 159, emphasis

original), style as much as content often being multilayered, shifting and ambiguous. Deconstruction is the tool employed to interpret such texts as definitive readings are proven to be impossible, or futile. As a result there can only be free-play within interpretation as evidenced in the deconstructive approach akin to (syn)aesthetic analysis (see also Derrida, 1976, 1978, 1981 and Barthes, 1975, 1982b, 1982f, 1987b, 1987c).
7. In terms of my own arguments for *play*text and the visceral impact of the written/spoken word in performance, Derrida asserts '[w]riting *represents* (in every sense of the word) enjoyment. It plays enjoyment, renders it present and absent. It is play' (1976: 312, emphasis original).
8. Henry Daniel identifies how '[t]he "remembrance" of physical actions...embodied in the body as bio-evolutionary is never lost' and can be 'set in motion through physical, mental, emotional and psychic...recognitive processes' (2000: 63–4). The body is able to remember 'the history of processes that it has undergone at the genetic and cultural level' which can encourage an entering into other's (and thus othered) realities (Daniel, 2000: 64). This recognition of corporeality allows for an immediacy within 'another kind of experiencing, a remembering or retracing of certain paths' (Daniel, 2000: 61). A simplified explanation is offered by Michael Taussig who refers to this as 'sense-data in the bank of the self' (1993: 98).
9. Taussig highlights, from an anthropological and ethnographic perspective, how all human origin histories expound slippage as central to the process. This 'slippage' is 'the attempt to trace the connection through history...of how one thing becomes another thing' in an evolutionary 'action of becoming different while remaining the same' (Taussig, 1993: 125). This foregrounds how humans already accept their inherent primordial connection through this sense of origin, or evolution, where human ancestry is understood to be scientifically, metaphorically or actually, of the earth. Thus, corporeal understanding can actuate a chthonic experience that reclaims this potential.
10. Intertextual, as I use it here, defines those creative works that employ a variety of 'texts' to produce and play with meaning. Kristeva's 'intertextuality' is a 'permutation of texts' (Kristeva, 1992e: 36, emphasis original), which foreshadows Horst Ruthrof's and Susan Broadhurst's intersemiotic analyses. Both argue that 'contemporary semiotics takes as its object *several semiotic practices* which it considers as *'translinguistic*; that is, they operate through and across language, while remaining irreducible to its categories as they are presently assigned'; 'verbal language cannot mean by itself but can do so only semiotically, i.e. in relation to and through corroboration by non-verbal systems' (Ruthrof, 1992: 6; see also Ruthrof, 1992, 1997 and Broadhurst, 1999a).
11. There has been scientific research into the connection between synaesthesia and the arts, such as Cretien van Campen's *The Hidden Sense: synaesthesia in art and science* (2008), John Harrison's *Synaesthesia the strangest thing* (2001: 115–40) and Cytowic's chapter in this area (2002: 295–320). These publications, located in the scientific world of neurocognitive research, provide case studies of synaesthesia as an influence in individual cases of artistic practice (Cytowic's interview with David Hockney is particularly engaging (2002: 312–19)). Cytowic also highlights how certain artworks are

'contrived synaesthesia' in that they are created to communicate the synaesthetic experience of the artist, such as Wassily Kandinsky's writings and colour expressions of emotion (2002: 319–20).
12. To quote Eugenio Barba, this focus on the slippage between the different languages of performance and appreciation in discursive terms becomes a pleasurable, 'struggle against the fixity of words' (1995: 141).
13. My consideration of these critical and performance theories is not intended to be exhaustive but instead to show how ideas within the theories under scrutiny support and elucidate my own argument for a (syn)aesthetic mode of production and appreciation. What should be clear is the delight that exists in the play and slippage between theories and schools of thought in order to elucidate my own theory for performance analysis.

1.1 Defining (syn)aesthetics

1. As Raymond Williams points out, 'aesthetics' in the Greek sense (and early nineteenth-century usage) defines the 'science' and conditions of sensuous perception. It is only latterly that the term has been adopted, in artistic practice in particular, to define in general terms the form and content of visual appearance and effect (see Williams, 1987: 31–2).
2. As opposed to the assumption of a mind/body split following René Descartes' philosophy known as Cartesian dualism. Here thought is independent of bodily experience so that the thinking mind, rather than the experiencing body, defines human beings. To concur with Elizabeth Grosz, I believe that '[o]nly when the relation between mind and body is adequately retheorized' will there be a valorization of 'the contributions of the body to the production of knowledge systems' (1994: 19). As Camille Paglia asserts, 'mind, which has enabled humanity to adapt and flourish as a species, has also infinitely complicated our functioning as physical beings' (1992: 16). My argument returns to the notion of the human body as a perceiving entity 'total and holistic, a completed and integrated system (albeit one that grows and transforms itself)' (Grosz, 1994: 13).
3. I use the term haptic (from the Greek, 'to lay hold of') alongside tactile as the latter tends to connote only the superficial quality of touch. Haptic, taken from Paul Rodaway's usage, emphasizes the tactile perceptual experience of the body as a whole (rather than merely the fingers) and also highlights the perceptive faculty of bodily kineasthesics (the body's locomotion in space). This encompasses the sensate experience of the individual's moving body, and the individual's perceptual experience of the moving bodies of others. Following this, within a live performance moment there comes about a 'reciprocity of the haptic system' of perception (Rodaway, 1994: 44) through this experience of tactile and kinaesthetic moments, whether actual or observed. Haptic reciprocity also lends itself to the 'Bayesian logic' of perception as discussed by K. Carrie-Armel and Ramachandran (2003). Their consideration of the experience of phantom limbs provides evidence of the brain's capacity for 'extracting statistical correlations in sensory input' which means that the body is an 'internal construct' that may be 'altered by the stimulus contingencies and correlations that one encounters'

(Carrie-Armel and Ramachandran, 2003: 1505–6). This supports the logic of sensation and experience that exists in the inner realm of the human body, which may manifest itself as external perception.

4. This play with the duality of the word 'sense' is fundamental to my argument for (syn)aesthetics. The term 'making-sense/*sense*-making' intends to clarify the fact that human perception, by its very definition, fuses 'the *reception of information* through the sense organs' with perception as '*mental insight*', that is, 'a sense made of a range of sensory information, with memories and expectations' (Rodaway, 1994: 10, emphasis original). Thus, perception as sensation, that is, corporeally mediated, and perception as cognition, intellectually mediated (accepting that the latter also involves cultural and social mediation) (see Rodaway, 1994: 11).

5. Cytowic interrogates a wealth of research in this area, especially the theories of Daphne Maurer and her colleagues (see 1988, 1993, 1996, 1999) to survey the diverse evidence in favour of the claim that all humans begin as synaesthetes. His findings support the validity of the argument, a key feature being that the limbic system, the area of the brain and the neurological impulses associated with sensual, emotional and cognitive memory, develops early in humans; 'all sensory inputs, external and visceral, must pass through the emotional limbic brain before being distributed to the cortex for analysis' (see Cytowic, 2002: 253, 271–93). Van Campen also provides fascinating evidence in this area (2008: 29–44). In terms of my focus on the chthonic in this book it is interesting to note van Campen's description of the multisensory nature of neonate and childhood perception as 'sensory primordial soup' (2008: 29). Cytowic's theories help to clarify what I intend by chthonic perception that is (re)instilled in humans by certain performance work as he observes, 'synesthesia is more mammalian than sapient ... not because it is somehow primitive but because the sensory percepts are closer to the essence of what it is to *perceive meaning* than are semantic abstractions' (2002: 10, emphasis original).

6. Those interested in dance may draw connections here with the movement qualities defined in Rudolf Laban's effort actions in his theory and methodology of Choreutics and Eukinetics. Interestingly Kane cites these directly, 'flash flicker burn wring press' and so on, in *4.48 Psychosis* (2000a: 29). Not only does the visceral-verbal play with the language denote visceral-psychological experience, Kane's ludic *play* with dance terminology also accentuates the exchange, or continuum, that exists between the visceral and analytical, the physical and the verbal in thought and practice.

7. This idea develops on a tactile and haptic level Peter Brook's idea of performance work which makes 'the invisible visible' (see Brook, 1986: 47).

8. Cytowic argues '[b]]ecause metaphor joins reason and imagination, the conceptual system which reality is based is in part imaginative' (2002: 279).

9. Merleau-Ponty also provides interesting arguments for the way in which language is conceived and experienced which connects with the visceral-verbal element of the (syn)aesthetic style (see 2005: 273–5 and 1974).

10. As Rebecca Schneider argues, when the body is prioritized in this way it demands a 'sensate involvement' from the audience (1997: 36).

11. I use 'cite' and 'citing' on a corporeal and sensate level to denote that which inscribes, marks, takes back to a source.

12. Taussig validates this notion, and confirms Scarry's 'sharability of sentience' (Scarry, 1985: 326), in stating that, sentient (re)presentation enables a 'flow[ing] into each others' otherness' (Taussig, 1993: 192). With such 'sentient contact', that draws on the 'tactile knowing of embodied knowledge', the work presented allows the perceiver to affect '[c]orporeal understanding' where 'you don't see as much as be hit' (Taussig, 1993: 30–1).
13. See Broadhurst and Machon, 2006 for an interrogation of live performance and various technologies including discussion of biotechnology as performance. Philip Auslander considers such 'shifting among realms' in the 'juxtaposition of the live and the digital' as a 'fusion' rather than a 'con-fusion, of realms' (1999: 38).
14. 'Difference' is a term adopted by a number of contemporary thinkers including Derrida, Barthes, Cixous and Irigaray. With Derrida's critical analysis and the significance of the 'a' in his *'différance'* (as in *différer*, 'to differ' and 'to defer') this has diverse meaning incorporating; the deferral of presence; the movement of difference in terms of oppositions, and the production of those differences (see Derrida, 1987a: 8–9). The wider understanding of 'difference' developed from the 'decentring' of human experience from a point where the 'existence of a norm or centre in all things was taken for granted...white Western [male] norms of dress, behaviour, architecture, intellectual outlook, and so on, provided a firm centre against which deviations, aberrations, variations could be detected and identified as "Other" and marginal' (Barry, 1995: 66–7). In the twentieth century these centres were eroded (Barry, 1995: 67) following the disruptions of World Wars, scientific progressions, intellectual and artistic revolutions and so on. Regarding Derridean *différance*, as a result of social, cultural and individual experience of this, there can be 'no absolutes or fixed points' instead all is ' "decentred" or inherently relativistic', rather than deviation from a given centre, 'all we have is "free play" ' (Barry, 1995: 66–7).

1.2 Connecting theories

1. Grosz states that, for Nietzsche, '[t]he body is the intimate and internal condition of all knowledges, especially of that knowledge which sees itself as a knowledge of knowledges – philosophy' (1994: 125). The Dionysian impulse in analysis is thus 'a bodily activity...capable of dynamizing and enhancing life' (Grosz, 1994: 128).
2. Walter F. Otto highlights how the mythic Dionysian possession was 'a stunning assault on the senses' (1965: 91), forcing individuals to give up the self to the intoxication of the body. The paradoxically sensual and disturbatory nature of this mythic Dionysian possession is 'startling, disquieting, violent' and 'arouses opposition and agitation', causing an ecstasy of disturbance and of wonder (Otto, 1965: 74).
3. Paglia highlights the importance of returning drama to the Dionysian in order to reconnect it with its latent primordial potential and to recapture its ritualistic and ludic potential (see Paglia, 1992). She asserts, 'Drama, a Dionysian mode, turned against Dionysus in making the passage from ritual to mimesis, that is, from action to representation' (1992: 6). For Paglia,

Western arts practice represses and evades the chthonian, 'earth's bowels, not its surface', in favour of an Apollinian aesthetic that revises 'this horror [of the chthonian] into imaginatively palpable form' (1992: 5–6). A chthonic practice of artistic disturbance rejects mimesis for ritual, for representation *within* action enabling a 'reconciliation' (Nietzsche, 1967a: 49) of humans with the primordial.

4. Here Nietzsche foreshadows the transgressive signification of Kristeva's *semiotic*, and Ruthrof and Broadhurst's intersemiotic approach. Ruthrof argues that Nietzsche sees linguistic analysis alone as presenting 'obstacles in our paths when we proceed to explore inner phenomena and impulses' (Nietzsche qtd in Ruthrof, 1992: 10). Broadhurst highlights the Dionysian 'commitment to immediacy ... a knowledge that does not proceed from analysis or concepts' (Broadhurst, 1999a: 31).

5. For the Formalists, defamiliarized language has the ability to 'make us *see* differently' by exercising 'a controlled violence upon practical language, which is thereby deformed in order to compel our attention to its constructed nature' (Selden, Widdowson and Brooker, 1997: 32). 'Practical' language is that which is used for 'acts of communication' functional and easily accessible, without the 'constructed quality' of the 'literary' (Selden, Widdowson and Brooker, 1997: 32). Defamiliarization, is also a term associated with Bertolt Brecht's Epic Theatre and his *Verfremdungseffekte* ('defamiliarization devices') whereby performance processes are made clear, designed to awaken an audience to an active and political way of receiving theatre, which resonates in particular in the discussion with Wallace and Kwei-Armah in Part 2.

6. Similarly, for Derrida festival, is, 'the movement of a birth, the continuous advent of *presence*' providing 'the moment of pure continuity' and 'the model of the continuous experience' (Derrida, 1976: 262–3, emphasis original) which establishes jouissance (the nearest translation being 'extreme pleasure' or 'unspeakable bliss'). Thus, jouissance is the experience of absolute presence or the moment of continuous presence, foregrounding an actualization of prae-sens and fused disturbing/exhilarating experience which clarifies Kant's 'negative pleasure' (1911: 91).

7. These ideas foreshadow notions of the writerly text. As Bakhtin asserts, '[a]rtistic form, correctly understood, does not shape already prepared and found content, but rather *permits content to be found and seen for the first time*' (1984: 43, emphasis added).

8. For Barthes, jouissance defines an experiential state of 'intense crisis' combining 'connotations of sexual orgasm and polysemic speech' (Selden, Widdowson and Brooker, 1997: 144). Stephen Heath translates Barthes' *plaisir* (pleasure) as, 'linked to cultural enjoyment and identity, to the cultural enjoyment of identity, to a homogenizing movement of the ego', which is contrasted with jouissance 'a radically violent pleasure' that 'shatters – dissipates, loses – that cultural identity, that ego' (Heath, 1987: 9).

9. '[Y]ou cannot speak "on" such a text, you can only speak "in" it, in its fashion' (Barthes, 1975: 22); 'Whenever I attempt to "analyze" a text which has given me pleasure, it is not my "subjectivity" I encounter but my "individuality", the given which makes my body separate from other bodies and appropriates its suffering or its pleasure: it is my body of [jouissance] I encounter' (Barthes, 1975: 62).

10. Leon S. Roudiez translates Kristeva's use of the term jouissance as 'totality of enjoyment' with a simultaneous sensual, sexual, spiritual, physical, conceptual and experiential capacity (1992: 16). It also 'implies the presence of meaning ... requiring it by going beyond it' via the phonic '*j'ouïs sens* = I heard meaning' (Roudiez, 1992: 16).
11. The genotext is in opposition to the 'phenotext' which is defined as, 'a structure', a matter of 'algebra' (Kristeva, 1999a: 121). The phenotext, includes the symbolic modality and denotes that language which 'obeys rules of communication' and is described linguistically in terms of 'competence' and 'performance' (Kristeva, 1999a: 121), thus, akin to the Formalists' 'practical language' (Selden, Widdowson and Brooker, 1997: 32). Kristeva's genotext is aligned with Barthes' pleasurable text, as Barthes states, 'writing aloud ... belongs to the geno-text' (Barthes, 1975: 66).
12. Like Barthes and Kristevas' theories, écriture féminine encompasses jouissance. Defined from a feminized, perspective, 'at the simplest level of meaning – metaphorical – woman's capacity for multiple orgasm indicates that she has the potential to attain something more than Total, something extra ... Real and unpresentable' (Cixous, 1993: 165–6). Phonically, as discussed above, with *j'ouïs sens* 'another level of activity is implied ... in which the word is all important' (Cixous, 1993: 165–6) and a sensate perceptual function, crucial.
13. Kristeva asserts that Artaud's theories argue for transgressive bodily signs which operate within and beneath language, exciting components of the chora via 'a dance which mobilises gestures' and voice (Kristeva, 2000: 268). Artaud's ideas collude with 'the traumatic and the archaic' and make manifest 'the unnameable place of the passions and drives, linked to the energies of the body' (Kristeva, 2000: 268). Edward Scheer asserts the 'vibrant ludic quality' to Artaud's work defining him as 'a practitioner of jouissance' (qtd. in Kristeva, 2000: 269).

1.3 (Syn)aesthetics in practice

1. The (syn)aesthetic hybrid develops Wagner's and the early Romantics arguments for the inherent unity of all the arts and embraces Artaud's arguments for a total theatre. Rose-Lee Goldberg affirms that such cross-fertilizing of various aesthetic disciplines, explored by Modernist artistic practitioners (Dadaists, Surrealists and so on), established the effectiveness of 'an exchange between the arts' in the pursuit of the 'development of a sensibility' (1996: 46, 9).
2. I use the term site-specific to describe those artistic events where the site is a key feature of the work itself. This is usually understood to be 'outside' of traditional performance venues (although with the case of Punchdrunk's *Masque of the Red Death* (2007–2008), the company took over London's Battersea Arts Centre, opened the whole building up to be played in, and completely transformed the internal structure with an Edgar Allen Poe inspired otherworld). 'Site-sympathetic' is a term chosen by Barrett as he feels this is more applicable to the nature of Punchdrunk's work where the company respond sensually to the site rather than placing a performance

within a given space. The site then defines, influences and shapes the work that is produced. With most site-based work the events designed would be reworked and become something other if translated to an alternative space. Nick Kaye's definition helps to clarify both terms: 'practices which, in one way or another, articulate exchanges between the work of art and the places in which its meanings are defined.... To move the site-specific work is to *re-place* it, to make it *something else*' (2001: 1–2, emphasis original). See also Hill and Paris, 2006 for interesting debate around performance and place.
3. Arguably, in the case of biotechnology the work may exist *as* the live performance. See Broadhurst and Machon, 2006 for further discussion of such practice.
4. Indeed in Part 2 Khan identifies how such an approach does not exist in the Kathak tradition; 'when I see a musician, I see dance, a dancer and I also see an actor. When I see a dancer I see a musician and an actor'.
5. See Foucault on arguments for the 'docile' human body as produced by social control (1991: 135–69).

Bibliography

Artaud, Antonin. 1971. *Collected Works: Volume Two*. Trans. Victor Corti. London: Calder & Boyars.
——. 1972. *Collected Works: Volume Three*. Trans. Alastair Hamilton. London: Calder & Boyars.
——. 1974. *Collected Works: Volume Four*. Trans. Victor Corti. London: Calder & Boyars.
——. 1978. *Collected Works: Volume One*. Trans. Victor Corti. London: John Calder.
——. 1993. *The Theatre and Its Double*. Trans. Victor Corti. London: Calder Publications Limited.
Auslander, Philip. 1999. *Liveness – Performance in a Mediatized Culture*. London and New York: Routledge.
Bakhtin, Mikhail. 1984. *Problems of Dostoevsky's Poetics*. Ed. and Trans. Caryl Emerson. (Theory and History of Literature, Vol. 8). Minneapolis, MN: University of Minnesota Press.
——. 2001. 'Rabelais and His World', *Performance Analysis – an Introductory Coursebook*. Eds Colin Counsell and Laurie Wolf. Trans. Hélène Iswolsky. London and New York: Routledge. 216–21.
Barba, Eugenio. 1995. *The Paper Canoe – a Guide to Theatre Anthropology*. Trans. Richard Fowler. London and New York: Routledge.
Barker, Howard. 1997. *Arguments for a Theatre*, 3rd edition. Manchester and New York: Manchester University Press.
——. 2001. Personal Interview. 30 May.
Barrett, Felix and Josephine Machon. 2007. 'Felix Barrett in Conversation with Josephine Machon', *Body, Space & Technology Journal*, Vol. 7, No. 1, Brunel University. http://people.brunel.ac.uk/bst/vol0701/home.html, accessed June 2008.
Barry, Peter. 1995. *Beginning Theory – an Introduction to Literary and Cultural Theory*. Manchester and New York: Manchester University Press.
Barthes, Roland. 1975. *The Pleasure of the Text*. Trans. Richard Miller. London: Jonathan Cape Ltd.
——. 1982a. 'A Lover's Discourse', *A Barthes Reader*. Ed. Susan Sontag. London: Jonathan Cape Ltd. 426–56.
——. 1982b. 'Authors and Writers', *A Barthes Reader*. Ed. Susan Sontag. London: Jonathan Cape Ltd. 185–93.
——. 1982c. 'Inaugural Lecture, College de France', *A Barthes Reader*. Ed. Susan Sontag. London: Jonathan Cape Ltd. 457–78.
——. 1982d. 'The Imagination of the Sign', *A Barthes Reader*. Ed. Susan Sontag. London: Jonathan Cape Ltd. 211–15.
——. 1982e. 'Roland Barthes by Roland Barthes', *A Barthes Reader*. Ed. Susan Sontag. London: Jonathan Cape Ltd. 415–25.
——. 1982f. 'Writing Degree Zero', *A Barthes Reader*. Ed. Susan Sontag. London: Jonathan Cape Ltd. 1982: 31–61.

Barthes, Roland. 1987a. 'From Work to Text', *Image – Music – Text*. Selected and translated by Stephen Heath. London: Fontana Press. 155–64.
———. 1987b. 'Lesson in Writing', *Image – Music – Text*. Selected and translated by Stephen Heath. London: Fontana Press. 170–8.
———. 1987c. 'The Death of the Author', *Image – Music – Text*. Selected and translated by Stephen Heath. London: Fontana Press. 142–8.
———. 1989. *Mythologies*. Trans. Annette Lavers. London: Paladin.
Bodies in Flight. 1996. *Do the Wild Thing! A Peepshow*. Choreographer/co-director Sara Giddens. Writer/co-director Simon Jones. Performers Jon Carnall, Jane Devoy and Dan Elloway. Composer Chris Austin. Design Bridget Mazzey. Music Elizabeth Whittam, Sarah Smale and Ben Rogerson. Artwork Jon Carnall.
———. 1998. *Constants*. Choreographer/co-director Sara Giddens. Writer/co-director Simon Jones. Composer Darren Bourne. Multimedia artist Caroline Rye. Performers Patricia Breatnach and Sheila Gilbert.
———. 1999–2000. *Deliver Us*. Choreographer/director Sara Giddens. Text/director Simon Jones. Performers Mark Adams and Polly Frame. Video Caroline Rye. Produced by Bodies in Flight at The Roadmender, Northampton. 15 June.
———. 2002. *Skinworks*. Choreographer/co-director Sara Giddens. Writer/co-director Simon Jones. Performers Polly Frame and Graeme Rose. Preview Bonington Gallery, Nottingham. 1 May.
———. 2004. *Who By Fire?* Choreographer/co-director Sara Giddens. Writer/co-director Simon Jones. Performers Polly Frame and Graeme Rose. Battersea Arts Centre, London.
———. 2007–8. *Model Love*. Choreographer/co-director Sara Giddens. Text/lyrics/co-director Simon Jones. Music and lyrics Sam Halmarack. Devised and performed by Catherine Dyson, Graeme Rose and Tom Wainwright. Photography Edward Dimsdale. Battersea Arts Centre, London.
Broadhurst, Susan. 1999a. *Liminal Acts: A Critical Overview of Contemporary Performance and Theory*. London: Cassell.
———. 1999b. 'The (Im)mediate Body: A Transvaluation of Corporeality', *Body & Society*, Vol. 5, No. 1, London, Thousand Oaks, CA and New Delhi: Sage Publications. 17–29.
———. 2007. *Digital Practice – Aesthetic and Neuroaesthetic Approaches to Performance and Technology*. Basingstoke and New York: Palgrave.
Broadhurst, Susan and Josephine Machon. (Eds). 2006. *Performance and Technology – Practices of Virtual Embodiment and Interactivity*. Basingstoke and New York: Palgrave.
Brook, Peter. 1986. *The Empty Space*. Middlesex: Pelican Books, Penguin.
Cannon, Glyn. 2004. *On Blindness*. London: Methuen.
Carnesky, Marisa. 1998. Writer/deviser/performer. *The Dragon Ladies – a Burlesque Revue*. Archive video. London, Raymond Revue Bar, Sunday 31 May and Sunday 12 April.
———. 1999a. Writer/deviser/designer/performer. *Jewess Tattooess*. Films by Alison Murray. Soundtrack Dave Knight with specially commissioned tracks by Katherine Gifford and James Johnson. Tattoos and set Alex Binnie. London, Battersea Arts Centre, 21 October.
———. 1999b. Writer/deviser/designer/performer. *Jewess Tattooess*. Films by Alison Murray. Soundtrack Dave Knight with specially commissioned tracks by

Katherine Gifford and James Johnson. Tattoos and set Alex Binnie. London, ICA, 9 December.

———. 2001. Writer/deviser/designer/performer. *Jewess Tattooess*. Films by Alison Murray. Copenhagen, Kanon Halleh.

———. 2002. Writer/deviser/designer/performer. *Carnesky's Burlesque Ghost Box*. Magic illusions. Paul Kieve. London. 21 June.

———. 2004. *Carnesky's Ghost Train*. Conception/director/designer/deviser/performer. Funded and commissioned by Arts Council England, Nesta, Hellhound, European Cultural Foundation, Warwick Arts Centre, Fierce, Mama Cash, Creative Lewisham Agency, Creative London, London Artists Projects. Brick Lane, London.

———. 2007. Writer/performer. *Magic War*. Dir. Flick Ferdinando. Dramaturg Lois Weaver. Design The Insect Circus. Lighting designer Sue Baynton. Co-performer Scott Waldrup. Soho Theatre, London. October–November.

———. 2008. *Carnesky's Ghost Train*. Conception/director/designer/deviser. In collaboration with Blackpool Illuminations and: Design Mark Copeland and Sarah Munro of Insect Circus. Illusions Paul Kieve, Dramatturg Natahsa Davis. Music Rohan Kriwaczek. Research Professor Vanessa Toulmin. Performers Geneva Foster Gluck, Empress Stah, Helen Plewis, Ashling Deeks, Rowan Fae, Ryan Styles, Amber Hickey, Agnes Czerna, Ruby Blues and Zamira Mummery. Blackpool.

Carrie-Armel, K. and V.S. Ramachandran. 2003. 'Projecting Sensations to External Objects: Evidence from Skin Conductance Response', *Proceedings B of The Royal Society*. (Proc. R. Soc. B). (270). June: 1499–506.

Churchill, Caryl. 1993. *Not Not Not Not Not Enough Oxygen and Other Plays*. Eds Terry Gifford and Gill Round. Essex: Longman.

———. 1994a. *The Skriker*. London: Nick Hern Books.

———. 1994b. Writer. *The Skriker*. Dir. Les Waters. Designer Annie Smart. Music Judith Weir. Movement Ian Spink. Lighting Christopher Toulmin. Cottosloe auditorium, Royal National Theatre, London.

———. 1997. *Blue Heart*. London: Nick Hern Books.

———. 1998a. Writer. *Blue Heart*. Dir. Max Stafford-Clark. Designer Julian McGowan. Pleasance Theatre, London. 9 March.

———. 1998b. 'The Lives of the Great Poisoners', *Plays: 3*. London: Nick Hern Books. 184–237.

———. 2000a. *Far Away*. London: Nick Hern Books.

———. 2000b. Writer. *Far Away*. Dir. Stephen Daldry. Performers Linda Bassett, Kevin McKidd, Anabelle Seymour-Julen and Katherine Tozer. Designer Ian McNeil. Royal Court Theatre Upstairs, London. 13 December.

———. Writer. 2001. *Far Away*. Dir. Stephen Daldry. Performers Linda Bassett, Kevin McKidd, Anabelle Seymour-Julen and Katherine Tozer. Designer Ian McNeil. Albery Theatre, London. 19 February.

———. 2002a. *A Number*. London: Nick Hern Books.

———. Writer. 2002b. *A Number*. Dir. Stephen Daldry. Designer Ian McNeil. Royal Court Jerwood Theatre Downstairs, London. 1 October.

———. 2002c. Text. *Plants and Ghosts*. Choreography Siobhan Davies. Movement material The Siobhan Davies Company dancers. Sound installation Max Eastley. Lighting Design Peter Mumford. Costumes Genevieve Bennett and Sasha Keir. Victoria Miro Gallery, London. 16 October.

Churchill, Caryl. 2006. *Drunk Enough to Say I Love You?* London: Nick Hern Books.
Churchill, Caryl and David Lan. 1998. 'A Mouthful of Birds', *Plays: 3. Caryl Churchill*. London: Nick Hern Books. 2–53.
Cixous, Hélène. 1993. *The Newly Born Woman. (Theory and History of Literature, Volume 24)*. With Catherine Clément. Trans. Betsy Wing. Minnesota: University of Minnesota Press.
——. 1995a. 'Aller á la mer', *Twentieth Century Theatre – a Sourcebook*. Ed. Richard Drain. Trans. Barbara Kerslake. London and New York: Routledge. 133–5.
——. 1995b. 'The Place of Crime, the Place of Pardon', *Twentieth Century Theatre – a Sourcebook*. Ed. Richard Drain. Trans. Eric Prenowitz. London and New York: Routledge. 340–4.
——. 1996. 'In October 1991 ...', *On the Feminine*. Ed. Mireille Calle. Trans. Catherine McGann. New Jersey: Humanities Press. 77–91.
Copeland, Roger. 1998. 'Between Description and Deconstruction', *The Routledge Dance Studies Reader*. Ed. Alexandra Carter. London and New York: Routledge. 98–107.
Curious. 1997. *The Day Don Came with the Fish*. Created and performed by Leslie Hill and Helen Paris. Commissioned by The London Filmmakers Coop and London Electronic Arts. The Lux, London, December 17–20.
——. 2000. *Deserter*. Written and performed by Leslie Hill and Helen Paris. Commissioned by the Project Arts Centre, Dublin.
——. 2000–7. *Vena Amoris*. Created by Leslie Hill and Helen Paris. Performed by Helen Paris. (Premiere, March 2000). Toynbee Studios, London. June 24.
——. 2003. *On the Scent*. Created and performed by Leslie Hill, Helen Paris and Lois Weaver. (Premiere at FIERCE! Birmingham June, 2003.) Toured London, North and South America, Europe, China and Australia.
——. 2004. *Essences of London*. Directed by Leslie Hill and Helen Paris. Filmed on location in the London boroughs of Brent, Hackney, Lambeth, Newham and Tower Hamlets.
——. 2005. *Lost & Found*. Created and performed by Leslie Hill, Helen Paris and Lois Weaver. Commissioned by The Public, West Bromwich and supported by Fierce Festival; Birmingham City Council (Urban Fusion programme); Bow Festival, London; British Council Artists Links, China and Arts Council England.
——. 2007a. *(be)longing*. Touring live performance and exhibitions. Written and performed by Leslie Hill and Helen Paris.
——. 2007b. *(be)longing*. Directed by Leslie Hill and Helen Paris. 35mm film.
——. 2008/9. *Autobiology*. Project created by Leslie Hill and Helen Paris. In collaboration with neuroscientist Professor Qasim Aziz.
Cytowic, Richard E. 1994. *The Man Who Tasted Shapes*. London: Abacus.
——. 1995. 'Synesthesia: Phenomenology and Neuropsychology – a Review of Current Knowledge', *Psyche – an Interdisciplinary Journal of Research on Consciousness*, Vol. 2, No. 10, July: 1–16. http://psyche.cs.monash.edu.au/v2/psyche-2-10-cytowic.html accessed December 2002.
——. 2002. *Synesthesia: A Union of the Senses*, 2nd edition. Cambridge MA: MIT Press.
Daniel, Henry. 2000. 'Re-Cognizing Corporeality', *Performing Processes – Creating Live Performance*. Ed. Roberta Mock. Bristol and Portland OR: Intellect Books. 61–8.

Deleuze, Gilles. 2004. *Francis Bacon – the Logic of Sensation*. Trans. Daniel W. Smith. London and New York: Continuum.
Deleuze, Gilles and Félix Guattari. 1999. *What Is Philosophy?* Trans. Graham Burchell and Hugh Tomlinson. London and New York: Verso.
Delmas, Gilles. 2008. Director. *bahok – Akram Khan Company/National Ballet of China – Letters sur le pont*. Ed. Marc Boyer. Artistic director, choreographer of bahok. Akram Khan. Music, Nitin Sawney. Commissioned by Akram Khan Dance Company.
Derrida, Jacques. 1976. *Of Grammatology*. Trans. Gayatri Chakravorty Spivak. Baltimore, MD and London: The Johns Hopkins University Press.
———. 1978. *Writing and Difference*. Trans. Alan Bass. London: Routledge.
———. 1981. *Dissemination*. Trans. Barbara Johnson. London: The Athlone Press.
———. 1987a. 'Implications: Interview with Henri Ronse', *Positions*. Translated and annotated by Alan Bass. London: The Athlone Press. 1–14, notes 97–8.
———. 1987b. 'Positions: Interview with Jean-Lois Houdebine and Guy Scarpetta', *Positions*. Translated and annotated by Alan Bass. London: The Athlone Press. 37–96, notes 99–114.
———. 1987c. 'Semiology and Grammatology: Interview with Julia Kristeva', *Positions*. Translated and annotated by Alan Bass. London: The Athlone Press. 15–36, notes 98–9.
Doyle, Maxine and Josephine Machon. 2007. 'Maxine Doyle in Conversation with Josephine Machon', *Body, Space & Technology Journal*, Vol. 7, No. 1, Brunel University. http://people.brunel.ac.uk/bst/vol0701/home.html, accessed June 2008.
Eichenbaum, Boris. 1965. 'The Theory of the Formal Method', *Russian Formalist Criticism: Four Essays*. Trans. and Intro. Lee T. Lemon and Marion J. Reis. Lincoln, NE and London: University of Nebraska Press. 99–139.
Eyre, Richard and Nicholas Wright. 2000. *Changing Stages – a View of British Theatre in the Twentieth Century*. London: Bloomsbury.
Foucault, Michel. 1991. *Discipline and Punish – the Birth of the Prison*. London: Penguin.
Galton, Francis. 1880a. 'Visualised Numerals', *Nature*, Vol. 21, No. 533: 252–6.
———. 1880b. 'Visualised Numerals', *Nature*, Vol. 22: 494–5.
———. 1907. *Inquiries into Human Faculty and Its Development*. London: Dent & Sons.
Goldberg, RoseLee. 1996. *Performance Art – from Futurism to the Present*. 1988. Singapore: World of Art – Thames and Hudson.
Grosz, Elizabeth. 1994. *Volatile Bodies – Toward a Corporeal Feminism*. Bloomington, IN and Indianapolis, IN: Indiana University Press.
Harris, Geraldine. 1999. *Staging Femininities – Performance and Performativity*. Manchester: Manchester University Press.
Harrison, John. 2001. *Synaesthesia the Strangest Thing*. Oxford: Oxford University Press.
Heath, Stephen. 1987. 'Translator's Note', *Image – Music – Text* by Roland Barthes. Selected and translated by Stephen Heath. London: Fontana Press. 7–11.
Hill, Leslie and Helen Paris (Eds). 2006. *Performance and Place*. Basingstoke and New York: Palgrave Macmillan.
Horton Fraleigh, Sondra. 1995. *Dance and the Lived Body*, New edition. Pittsburgh, PA: University of Pittsburgh Press.
Irigaray, Luce. 1985. *Speculum of the Other Woman*. Trans. Gillian C. Gill. Ithaca, NY: Cornell University Press.

Irigaray, Luce. 1991. *Marine Lover of Friedrich Nietzsche*. Trans. Gillian C. Gill. New York: Columbia University Press.
——. 1999a. 'The Bodily Encounter with the Mother', Trans. David Macey. *The Irigaray Reader*. (1991). Ed. Margaret Whitford. Oxford: Blackwell. 34–46.
——. 1999b. 'The Power of Discourse and the Subordination of the Feminine', Trans. Catherine Porter with Carolyn Burke. *The Irigaray Reader*. (1991). Ed. Margaret Whitford. Oxford: Blackwell. 118–32.
——. 1999c. 'Questions', Trans. Catherine Porter with Carolyn Burke. *The Irigaray Reader*. (1991). Ed. Margaret Whitford. Oxford: Blackwell. 133–9.
——. 1999d. 'The Three *Genres*', Trans. David Macey. *The Irigaray Reader*. (1991) Ed. Margaret Whitford. Oxford: Blackwell. 140–53.
——. 1999e. 'Volume without Contours', Trans. David Macey. *The Irigaray Reader*. (1991) Ed. Margaret Whitford. Oxford: Blackwell. 53–67
——. 1999f. 'When Our Lips Speak Together', *Feminist Theory and the Body – a Reader*. Eds Janet Price and Margrit Shildrick. Edinburgh: Edinburgh University Press. 82–90.
Kane, Sarah. 1996. *Blasted* and *Phaedra's Love*. London: Methuen.
——. 1998a. *Cleansed*. London: Methuen.
——. 1998b. *Crave*. London: Methuen.
——. 1998c. Writer. *Crave*. Dir. Vicky Featherstone. Designer Georgia Sion. Royal Court Theatre at The New Ambassadors, London. 28 September.
——. 2000a. *4.48 Psychosis*. London: Methuen.
——. 2000b. Writer. *4.48 Psychosis*. By Sarah Kane. Dir. James Macdonald. Designer Jeremy Herbert. Royal Court Theatre Upstairs, London. 1 July.
——. 2001a. Writer. *Blasted*. (Revival.) By Sarah Kane. Dir. James Macdonald. Designer Hildegard Bechtler. Royal Court Theatre, London. 11 April.
——. 2001b. Writer. *Cleansed*. (Rehearsed reading.) By Sarah Kane. Dir. James Macdonald. Royal Court Theatre, London. 11 April.
——. 2001c. Writer. *Crave*. (Revival.) By Sarah Kane. Dir. Vicky Featherstone. Designer Georgia Sion. Royal Court Theatre, London. 6 June.
——. 2001d. Writer. *4.48 Psychosis*. (Revival.) By Sarah Kane. Dir. James Macdonald. Designer Jeremy Herbert. Royal Court Theatre, London. 6 June.
——. 2007. Writer. *Blasted*. Dir. Jenny Sealey. Performers Jennifer Jay, Gerard McDermott and David Toole. Design Jenny Sealey and Jo Paul. Graeae at the Soho Theatre. London. 16 January–3 February.
Kant, Immanuel. 1911. *Kant's Critique of Human Judgement*. Trans., introductory essays, notes and analytical index James Creed Meredith. Oxford: Clarendon Press.
——. 1978. *Critique of Judgement*. Trans. James Meredith. Oxford: Clarendon Press.
Kaye, Nick. 2001. *Site-Specific Art – Performance, Place and Documentation*. London and New York: Routledge.
Khan, Akram. 2000–2. *Rush*. Choreographer/Performer/Costume designer. Composer. Andy Cowton. Lighting Designer. Michael Hulls. Dancers. Moya Michael. Gwyn Emberton, Inn Pang Ooi. Premiere, Midlands Arts Centre, Birmingham October 2000.
——. 2002–3. *Kaash*. Artistic Director/Choreographer/Performer. Composer Nitin Sawhney Additional Music 'Spectre' by John Oswald played by The Kronos Quartet. Set Designer Anish Kapoor. Light Designer Aideen Malone. Costumes Designer Saeunn Huld. Dancers Rachel Krische, Moya Michael, Inn Pang Ooi,

Shanell Winlock and Eulalia Ayguade Farro. Premiere: 28 March 2002, Creteil, France.

———. 2004–5. *ma*. Artistic director, Choreographer/Performer. Composer Riccardo Nova. Songs by Faheem Mazhar. Lighting Designer Mikki Kunttu. Set Designer illur malus islandus. Text Hanif Kureishi. Dramaturgy Carmen Mehnert. Dancers Eulalia Ayguade Farro, Anton Lachky, Navala Chaudhari, Young Jin Kim, Duan Ni Nikoleta Rafaelisova and Shanell Winlock.

———. 2005. Choreographer/performer. *Zero Degrees*. Co-Choreographer/performer Sidi Larbi. Design/environment and sculpture Antony Gormley. Composer Nitin Sawhney. Sadlers Wells. London. July.

———. 2006. Choreographer/performer. *Sacred Monsters*. Co-Performer. Sylvie Guillem. Sadlers Wells, London. September.

———. 2008a. Choreographer. *bahok*. With the National Ballet of China. Material devised and performed by Eulalia Ayguade Farro, Kim Young Gin, Meng Ningning, Andrej Petrovic, Saju, wang Yitong, Shanell winlock and Zhang Zhenxin. Sadlers Wells, London. June.

———. 2008b. Choreographer/performer. *in-i*. Co-Director/performer Juliette Binoche. Visual Design Anish Kapoor. Composer Philip Sheppard. Lighting Design Michael Hulls. Dramaturg Guy Cools. Lyttleton Theatre, The National. London. September 9.

Kristeva, Julia. 1982. *Powers of Horror – an Essay on Abjection*. Trans. Leon S. Roudiez. New York: Columbia University Press.

———. 1991. *Strangers to Ourselves*. Trans. Leon S. Roudiez. New York: Columbia University Press.

———. 1992a. 'From One Identity To Another', *Desire in Language – a Semiotic Approach to Literature and Art*. Ed. Leon S. Roudiez. Trans. Thomas Gora, Alice Jardine and Leon S. Roudiez. Oxford: Blackwell. 124–47.

———. 1992b. 'Giotto's Joy', *Desire in Language – a Semiotic Approach to Literature and Art*. Ed. Leon S. Roudiez. Trans. Thomas Gora, Alice Jardine and Leon S. Roudiez. Oxford: Blackwell. 210–36.

———. 1992c. 'How Does One Speak to Literature?', *Desire in Language – a Semiotic Approach to Literature and Art*. Ed. Leon S. Roudiez. Trans. Thomas Gora, Alice Jardine and Leon S. Roudiez. Oxford: Blackwell. 92–123.

———. 1992d. Preface. *Desire in Language – a Semiotic Approach to Literature and Art*. Ed. Leon S. Roudiez. Trans. Thomas Gora, Alice Jardine and Leon S. Roudiez. Oxford: Blackwell. vii–xi.

———. 1992e. 'The Bounded Text', *Desire in Language – a Semiotic Approach to Literature and Art*. Ed. Leon S. Roudiez. Trans. Thomas Gora, Alice Jardine and Leon S. Roudiez. Oxford: Blackwell. 36–63.

———. 1992f. 'The Novel as Polylogue', *Desire in Language – a Semiotic Approach to Literature and Art*. Ed. Leon S. Roudiez. Trans. Thomas Gora, Alice Jardine and Leon S. Roudiez. Oxford: Blackwell. 159–209.

———. 1999a. 'Revolution in Poetic Language', Trans. Margaret Waller. *The Kristeva Reader*. Ed. Toril Moi. Oxford: Blackwell. 89–136.

———. 1999b. 'The System and the Speaking Subject', Trans. Alice Jardine, Thomas Gora and Leon S. Roudiez. *The Kristeva Reader*. Ed. Toril Moi. Oxford: Blackwell. 24–33.

———. 1999c. 'Women's Time', Trans. Alice Jardine and Harry Blake, *The Kristeva Reader*. Ed. Toril Moi. Oxford: Blackwell. 187–213.

———. 2000. 'Artaud: Madness and Revolution – Interview with Julia Kristeva', *100 Years of Cruelty – Essays on Artaud*. Ed. Edward Scheer. Interview by Edward Scheer, conducted by Shan Benson. Sydney: Power Publications and Artspace. 263–78.

Lepage, Robert. 1993. Director/performer. *Needles and Opium*. Lyttleton Theatre at The National Theatre, London.

———. 1997a. *Connecting Flights*. With Rémy Charest. Trans. Wanda Romer Taylor. London: Methuen.

———. 1997b. 'Robert Lepage in discussion with Richard Eyre', *The Twentieth Century Performance Reader*. Eds Micheal Huxley and Noel Witts. London and New York: Routledge. 237–47.

Luria, A.R. 1969. *The Mind of a Mnemonist*. Trans. Lynn Solotaroff. London: Jonathan Cape Ltd.

Machon, Josephine. 2001a. '(Syn)aesthetics and Disturbance – a Preliminary Overview', *Body, Space, & Technology*, Vol. 1, No. 2, Brunel University Centre for Contemporary and Digital Performance. http://people.brunel.ac.uk/bst/vol0102/index.html accessed January 2009.

———. 2001b. 'to Deliver Us from (Syn)aesthetics', Flesh & Text: Far Ahead Publications. Interactive CD-ROM article

———. 2007. 'The (Syn)aesthetics of Punchdrunk's Site-sympathetic Work', *Body, Space & Technology Journal*, Vol. 7, No. 1, Brunel University. http://people.brunel.ac.uk/bst/vol0701/home.html, accessed June 2008.

Maurer, Daphne. 1993. 'Neonatal Synesthesia: Implications for the Processing of Speech and Faces', *Developmental Neurocognition: Speech and Face Processing in the First Year of Life*. Eds D de Boysson-Bardies et al. Dordecht: Kluwer. 109–24.

Maurer, Daphne and C. Maurer. 1988. *The World of the Newborn*. New York: Basic Books.

Maurer, Daphne and C.J. Mondlach. 1996. 'Synesthesia: A Stage of Normal Infancy?', *Proceedings of the Twelfth Annual Meeting of the International Society of Psychophysics*. Ed. S.C. Masin. Padua: The International Society of Psychophysics. 107–12.

Maurer, Daphne, C.L. Stager and C.J. Mondlach. 1999. 'Cross-Modal Transfer of Shape Is Difficult to Demonstrate in One-Month-Olds', *Child Development*, 70: 1047–57.

Merleau-Ponty, Maurice. 1964. *The Primacy of Perception*. Ed. James M. Edie. Trans. James M. Edie et al. USA: Northwestern University Press.

———. 1974. *The Prose of the World*. Ed. Claude Lefort. Trans. John O'Neill. London: Heinemann Books.

———. 2005. *The Phenomenology of Perception*. Trans. Colin Smith. London and New York: Routledge.

Nietzsche, Friedrich. 1967a. *The Birth of Tragedy and the Case of Wagner*. Trans. Walter Kaufmann. New York: Vintage Books.

———. 1967b. *Thus Spake Zarathustra*. Trans. Thomas Common, revised by Oscar Levy and John L. Beevers. London: George Allen and Unwin Ltd.

———. 1968. *The Will to Power*. Trans. Walter Kaufmann and R.J. Hollingdale. Ed. Walter Kaufmann. New York: Vintage.

———. 1969. *On the Genealogy of Morals*. Trans. Walter Kaufmann and R.J. Hollingdale; and *Ecce Homo*. Trans. and Ed. Walter Kaufmann. New York: Vintage.

———. 1994. *Human, All Too Human*. Trans. Marion Faber and Stephen Lehmann. London: Penguin.
Novarina, Valère. 1993. 'Letter to the Actors'. *The Drama Review*. Trans. Allen S. Weiss. Vol. 37, No. 2 (T138), Summer: 95–118.
———. 1996. *The Theater of the Ears*. Trans., Ed. and Intro. Allen S. Weiss. Los Angeles, CA: Sun & Moon Press.
Otto, Walter. F. 1965. *Dionysus Myth and Cult*. Trans. Robert B. Palmer. Indiana, IN: Indiana University Press.
Paglia, Camille. 1990. *Sexual Personae – Art and Decadence from Nefertiti to Emily Dickinson*. London: Penguin, 1992.
Paglia, Camille. *Sexual Personae – Art and Decadence from Nefertiti to Emily Dickinson*. London: Penguin.
Paris, Helen. 2006. 'Too Close for Comfort: One-to-One Performance', *Performance and Place*. Eds Leslie Hill and Helen Paris. Basingstoke and New York: Palgrave Macmillan. 179–91.
Phelan, Peggy. 1993. *Unmarked – the Politics of Performance*. London and New York: Routledge.
———. 1997. *Mourning Sex – Performing Public Memories*. London and New York: Routledge.
———. 1998. Intro., 'The Ends of Performance', *The Ends of Performance*. Eds Peggy Phelan and Jill Lane. New York and London: New York University Press. 1–19.
Phelan, Peggy and Jill Lane. (Eds). 1998. *The Ends of Performance*. New York and London: New York University Press.
Punchdrunk. 2003–4. *Sleep No More*. Dir. Felix Barrett. Choreographer Maxine Doyle. The Beaufoy Building, Kennington, London. November 2003–January 2004.
———. 2005. *The Firebird Ball*. Dir. Felix Barrett. Choreographer Maxine Doyle. Offley Works. London. January–March 2005.
———. 2006–7. *Faust*. Dir. Felix Barrett. Choreographer Maxine Doyle. Producer Colin Marsh. In association with the National Theatre. 21 Wapping Lane, London. October 2006–March 2007.
———. 2007–8. *The Masque of the Red Death*. Dir. Felix Barrett. Choreographer Maxine Doyle. Producer Colin Marsh. In association with the National Theatre. Battersea Arts Centre. September 2007–April 2008.
Ramachandran, V.S. and E.M. Hubbard. 2001a. 'Psychophysical Investigations into the Neural Basis of Synaesthesia', *Proceedings of the Royal Society*, 268: 979–83.
———. 2001b. 'Synaesthesia – a Window into Perception, Thought and Language', *Journal of Consciousness Studies*, Vol. 8, No. 12: 3–34.
———. 2003. 'Hearing Colors, Tasting Shapes', *Scientific American*, Vol. 288, No. 5, May: 52–9.
———. 2005. 'Neurocognitive Mechanisms of Synesthesia', *Neuron*, Vol. 48, No. 3, November: 509–20.
Rodaway, Paul. 1994. *Sensuous Geographies – Body, Sense and Place*. Routledge: London and New York.
Roudiez, Leon S. 1992. Intro., *Desire in Language – a Semiotic Approach to Literature and Art*. By Julia Kristeva. Ed. Leon S. Roudiez. Trans. Thomas Gora, Alice Jardine and Leon S. Roudiez. Oxford: Blackwell. 1–20.
Ruthrof, Horst. 1992. *Pandora and Occam: On the Limits of Language and Literature*. Bloomington, IN: Indiana University Press.

Ruthrof, Horst. 1995. 'Meaning: An Intersemiotic Perspective', *Semiotica*, Vol. 104, No. (1/2): 23-43.

Ruthrof, Horst. 1997. *Semantics and the Body: Meaning from Frege to the Postmodern*. Toronto: University of Toronto Press.

Scarry, Elaine. 1985. *The Body in Pain – the Making and Unmaking of the World*. New York and Oxford: Oxford University Press.

Scheer, Edward. (Ed.) 2000. *100 Years of Cruelty – Essays on Artaud*. Sydney: Power Publications and Artspace.

Schneider, Rebecca. 1997. *The Explicit Body in Performance*. London: Routledge.

Sealey, Jenny. 2004. Director. *Bent*. Written by Martin Sherman. London, Cochrane Theatre, 14-23 October.

———. 2002. Director. *Peeling*. Written by Kaite O' Reilly. London, Soho Theatre. 3-13 April.

———. 2004. Co-director. *On Blindness*. Writer Glyn Cannon. Co-directors Vicky Featherstone and Steve Hoggett. Performers Mat Fraser, Scott Graham, Steve Hoggett, Karina Jones, Jo McInnes and David Sands. Designer Julian Crouch. Soundscore Nick Powell. London, Soho Theatre.

———. 2007. Director. *Blasted*. Written by Sarah Kane. London, Soho Theatre, January.

———. 2008. Director. *Static*. Written by Dan Rebellato. London, Soho Theatre, 22 April-10 May.

Selden, Raman, Peter Widdowson and Peter Brooker. 1997. *A Reader's Guide to Contemporary Literary Theory*, 4th edition. London: Prentice Hall Harvester Wheatsheaf.

Shklovsky, Victor. 1965. 'Art as Technique', *Russian Formalist Criticism: Four Essays*. Trans. and Intro. Lee T. Lemon and Marion J. Reis. Lincoln, NE and London: University of Nebraska Press: 3-24.

Shunt Theatre Collective. 1998. *Twist*. Created and performed by The Shunt Collective. Battersea Arts Centre, London. October.

———. 1999-2001. *The Ballad of Bobby François*. Created and performed by The Shunt Collective. Arch 12a, Bethnal Green, May and October (1999). Pleasance Theatre, Edinburgh, (2000), The Drome, London, January (2001).

———. 2001. *The Tennis Show*. Created and performed by The Shunt Collective. The Bargehouse, London. November.

———. 2002-3. *Dance Bear Dance*. Created and performed by Shunt. Arch 12a Bethnal Green, London. May 2002-August 2003.

———. 2004-5. *Tropicana*. In collaboration with the National Theatre. Created by Shunt in collaboration with Nigel Barrett, Julie Boules, Sarah Cant, Suzanne Dietz, David Farley, Geneva Foster-Gluck, Helena Hunter, Leila Jones, Simon Kane, Paul Mari, Slivia Mecuriali, Ben Ringham, Max Ringham and Chris Teckkam. Sound and music in collaboration with Conspiracy. The Shunt Vaults, London. September-July.

———. 2006. *Amato Saltone*. In collaboration with the National Theatre. Created by Shunt. The Shunt Vaults, London.

Sierz, Aleks. 2001. *In-Yer-Face Theatre – British Drama Today*. London: Faber and Faber.

Sontag, Susan. 1982a. *A Susan Sontag Reader*. Middlesex: Penguin.

———. 1982b. 'Writing Itself: On Roland Barthes', *A Barthes Reader*. Ed. Susan Sontag. London: Jonathan Cape Ltd. vii-xxxviii.

Taussig, Michael. 1993. *Mimesis and Alterity – a Particular History of the Senses*. New York and London: Routledge.
Tozer, Kathy. 2001. (Older Joan, *Far Away*. Royal Court Theatre Upstairs, London. 13 December and Albery Theatre, London. 19 February, 2001). Personal Interview. 12 January: 1–16.
Turner, Victor. 1982. *From Ritual to Theatre – the Human Seriousness of Play*. New York: Performing Arts Journal Publications.
———. 1986. *The Anthropology of Performance*. New York: PAJ Publications.
Van Campen, Cretien. 2008. *The Hidden Sense – Synaesthesia in Art and Science*. Cambridge, MA: MIT Press.
Wallace, Naomi. 2002. *In the Heart of America and Other Plays*. New York: Theatre Communications Group.
———. 2007a. *Things of Dry Hours*. London: Faber & Faber.
———. 2007b. Writer. *Things of Dry Hours*. Dir. Kwame Kwei-Armah. Performers Steven Cole Hughes, Erika LaVonn and Roger Robinson. A Centerstage Production. Baltimore, USA. April–June.
Ward, Nigel. 1999. 'Twelve of the Fifty-One Shocks of Antonin Artaud', *New Theatre Quarterly*, Vol. XV, Part 2 (NTQ 58). Cambridge: Cambridge University Press. May: 123–30.
Weiss, Allen S. 1993. 'Mouths of disquietude – Valère Novarina Between the Theatre of Cruelty and Écrit Bruts', *The Drama Review*, Vol. 37, No. 2 (T138), Summer: 80–94.
Williams, Raymond. 1987. *Keywords – a Vocabulary of Culture and Society*. London: Flamingo.

Websites

Akram Khan Company: http://www.akramkhancompany.net/
Bodies in Flight: http://www.bodiesinflight.co.uk/
Curious: http://www.placelessness.com/
Graeae: http://www.graeae.org/
Marisa Carnesky: http://www.carnesky.com/ and http://carneskysghosttrain.com/
Punchdrunk: http://www.punchdrunk.org.uk/
Shunt Theatre Collective: http://www.shunt.co.uk/

Index

Anderson, Laurie, 30
Artaud, Antonin, 1, 8, 13, 34, **44–6**, 47, 48, 49, 80, 135, 203
  and the senses, 45
  and text, 73, 74, 75
  Theatre of Cruelty, **44–6**, 49
  total theatre, 55, 56, 65, 203
  'writing *of* the body', 44, 62, 63, 65, 68
  see also Derrida, Jacques

Bakhtin, Mikhail, 37, **38**, 202
  carnival, carnivalization, 37, **38**, 73, 47, 56. 64, 79, 124
  'polyphonic consciousnesses', 38, 64, 70, 86
  see also Russian Formalists, the
Barker, Howard, 8, 34, 48, **48–50**, 51, 74, 78
  and the audience, 48, 49, 72
  and imagination, 48, 49
  Theatre of Catastrophe, **48–50**
Barrett, Felix, 25, 85, **89–99**, 203
  see also Punchdrunk
Barthes, Roland, 8, 34, **39–40**, 47, 74, 197–8, 201
  'anterior immediacy', 40
  disfiguration of language, 39, 78
  and *jouissance*, **39–40**, 46, 202, 203
  pleasurable texts, 34, **39–40**, 43, 203
  writing aloud, 40, 46, 79, 203
Bassett, Linda, 7, 19, 22, 86, **144–52**
Bausch, Pina, 2, 114
  Bluebeard, 3
  Tanztheater Wuppertal, 29
Beckett, Samuel, 1, 32
  Not I, 3
Berkoff, Steven, 3
  Metamorphosis, 3
biotechnologies, 25, 50, 59, 188–90, 201, 204

Bodies in Flight, 7, 25, 29, 34, 55, 56, 59, 61, 63, 65, 66, 67, 71, 79, 87, **172–83**
body, 1–2, 3–4, 6, 9, 13, 14, 19, 20, **21–4**, 26, 27–8, 60, 87, 197
  as cite, 23–4, 28, 42, 45, 55, 59, 64, 66, 68, 86, 103–4, 109–10, 112–23, 158, 188–9, 192, 195, 198, 199, 200
  the performing body, **62–9**, 86, 89–99, 175–81
  as site and source, 42, 58–9, 68, 86, 112–23, 124–31, 189–90
  and text, 69–80, 86, 132–43, 169, 175–7, 190
  and theory, 34–53, 201, 202, 203, 204
Brecht, Bertolt, 135, 202
  Epic Theatre, 202
  Verfremdungseffekte, 202
Brik, Osip, 38
  and rhythm, 78
  see also Russian Formalists, the
Broadhurst, Susan, 5, 8, 34, 42, **50–2**, 69, 72, 197
  and the body, 62, 64, 65, 66, 67, 68
  Derridean 'jarring metaphors', 50, 52, 59
  and digital technologies, 50, 52
  and hybridity, 51, 59, 61
  intersemiotic analysis, 6, 50, 52, 198, 202
  see also Ruthrof, Horst
  the Liminal, 50–2, 59
  and Machon, Josephine, 201, 204
  and Turner, Victor, 50
Brook, Peter, 123, 200
  Theatre Bouffes du Nord, 29
Brooks, Pete, 30

217

Cannon, Glyn, 61, 71, 79, 87, **160–71**
Carnesky, Marisa, 5, 7, 23, 30, 55, 56, 58, 59, 60, 61, 64, 65, 67, 79, 86, **124–31**, 195
  Ghost Train, 58, 60, 126, 128–31
  Jewess Tattooess, 58–9, 65, 125–8
  Magic War, 126–8
Cartesian mind/body split, 46, 189, 199
Churchill, Caryl, 7, 19, 24, 31, 32, 42, 55, 64, 69, 73, 74, 79, **144–52**
  Blue Heart, 78, 149
  Drunk Enough To Say I Love You?, 78
  Far Away, 71, 72, 74, 78, 86, **144–52**
  The Lives of the Great Poisoners, 71
  A Number, 78
  The Skriker, 3, 67, 71, 144, 145, 148–9, 197
Cixous, Hélène, 8, 28, 34, **42–4**, 46, 47, 63, 197, 201, 203
  see also écriture féminine
Clachan, Lizzie, 25, 54, 58, 85, **100–11**
Curious, 7, 25, 29, 55, 56, 59, 61, 63, 65, 67, 79, 87, **184–96**
  Autobiology, 63, 187–90, 191
  The Day Don Came With The Fish, 190, 192
  Deserter, 186
  On The Scent, 63, 184, 187, 189, 191, 193, 194
  Vena Amoris, 66, 185–7
  and the 'visceral-virtual', 59, 184, 194
  see also Hill, Leslie; Paris, Helen
Cytowic, Richard E., 6, **15–20**, 32, 198, 200

Daldry, Stephen, 148
Dance Theatre, 29
defamiliarization, 37, 78, 202
  see also Shklovsky, Victor, *Ostranenie*
De La Guarda, 3, 30
  Villa! Villa!, 3
Derrida, Jacques, 8, 52, 55, 62, 65, 197, 198, 201, 202
  deconstruction, 198

différance, 201
'wide jarring metaphors', 50, 52, 59
  see also Broadhurst, Susan
'writing of the body' 44, 62, 63, 65, 68
  see also Artaud, Antonin
Doyle, Maxine, 25, 58, 68, 85, 87, **89–99**
  see also Punchdrunk
DV8, 3
  Dead Dreams of Monochrome Men, 3
  Lloyd Newson, 113

écriture féminine, 28, 34, **42–4**, 63, 203
  and the body, 43, 63
  and *jouissance*, 203
  and marginal experience, 63
  and the primordial, 43
  and the unconscious, 44
  see also Cixous, Hélène; Irigaray, Luce

feminized practice, **26–9**, 70
  and the body, 27–8
  and theory, 28–9, 203
  see also écriture féminine

Galton, Francis, 15
gesamtkunstwerk, 4, 44, 51
  defined, 4
Giddens, Sara, 79, 87, **172–83**
Graeae, 7, 25, 29, 55, 59, 60, 61, 64, 67, 79, 87, **160–71**
  Bent, 164, 170
  Blasted, 161, 162
  On Blindness, 71, 79, 87, **160–71**
  Peeling, 161, 164
Greenaway, Peter, 5, 192

haptic, 14, 17, 55, 56, 57, 58, 60, 62, 67, 69, 87, 98, 199, 200
  defined, 199
Harris, Geraldine, 28, 197
Harrison, John, 15, 198
Hill, Leslie, 5, 25, 26, 63, **184–96**, 204
  see also Curious
Hockney, David, 198

Horton-Fraleigh, Sondra, 2, 197
hypermnesis, 18, 21
  defined, 18

interdisciplinary practice, 2, 3, 5, 9, 14, **29–30**, 53, **55–62**, 71, 95, 164
  and dance theatre, 29
  intercultural performance, 2, 29–30
  intersemiotic analysis, 6, 50, 52, 198, 202
  see also Broadhurst, Susan; Ruthrof, Horst
intertextual, intertextuality, 6, 14, 30, 35, 41, 42, 60, 80, 198
  see also Kristeva, Julia
Ionesco, Eugene, 32
Irigaray, Luce, 8, 28, 34, **42–4**, 47, 63, 71, 78, 197, 201
  see also écriture féminine

Jarman, Derek, 5
Jodorowsky, Alejandro, 5, 130
  Santa Sangre, 130
Jones, Simon, 56, 71, 79, 87, **172–83**
jouissance, 39, 41, 45, 46, 203
  defined, 39, 202, 203
  see also Barthes, Roland; écriture féminine; Kristeva, Julia

Kandinsky, Wassily, 199
Kane, Sarah, 7, 24, 31, 32, 42, 55, 64, 67, 69, 73, 74, 79, **153–9**, 163, 200
  4.48 Psychosis, 61, 73, 74, 86, **153–9**, 200
  Blasted, 73, 161, 162, 163
  Cleansed, 123
  Crave, 157
Kant, Immanuel, 5, 8, **21–2**, 39, 49, 202
  'free-play' of imagination, 5, 28, 30, 32, 75, 198
  negative pleasure, 21, 22, 39, 49, 202
  the sublime, 21–2
Khan, Akram, 7, 29, 55, 61, 64, 67, 68, 79, 86, **112–23**, 195, 204

Akram Khan Dance Company, 29, 76, 77, 86
  bahok, 65, 114, 116, 118, 119, 121, 123
  and Cherkaoui, Sidi Larbi, 65, 76
  in-i, 65, 120, 121, 123
  Kaash, 113, 114, 115, 116
  Ma, 114, 121
  Sacred Monsters, 65, 119, 122
  Zero Degrees, 65, 75, 76, 77, 116, 117, 120, 123
Kotting, Andrew, 5, 192
  Filthy Earth, 192
Kristeva, Julia, 8, 34, **40–2**, 43, 44, 47, 65, 197, 198, 203
  genotext, 40, 41, 42, 43, 203
  and the instinctual body, 41
  and *jouissance*, 203
  semiotic *chora*, 44, 47, 48, 67, 202, 203
  symbolic, 41, 42, 203
Kwei-Armah, Kwame, 7, 22, 24, 70, 79, 86, **132–43**, 202

Laban, Rudolf, 200
Lepage, Robert, xviii, 2, 85
  Ex Machina, 29, 30
  Needles & Opium, xviii, 2
  live performance, 24–6
  and technology, 25–6
Luria, A.R., 6, 15, 16, **17–18**, 19, 20, 47, 49, 78
Lynch, David, 5

Macdonald, James, 153
McInnes, Jo, 7, 22, 24, 61, 79, 86, **153–9**, 167, 168
Merleau-Ponty, Maurice, 8, 21, 22–3, 52, 200

New Writing, **31–2**
Nietzsche, Friedrich, 8, 22, 34, **35–7**, 42, 43, 51, 64, 201, 202
  Apollinian, 35, 36, 37, 41
  Dionysian, 34, **35–7**, 38, 41, 42, 43, 44, 45, 48, 50, 51, 55, 64, 74, 78, 201, 202
  eternal recurrence, 43
  'resonance', 36
  'special memory', 36

Novarina, Valère, 8, 34, **46–8**, 69, 78, 79, 80
  and the body, 46–7
  'the speak', 46
  'text that dances', 46
  Theatre of the Ears, 46–8

Paris, Helen, 5, 25, 26, 59, 79, 87, **184–96**, 204
  see also Curious
Phelan, Peggy, 1, 2, 25, 185
presence, 21–6
  and *prae-sens*, 25, 39, 40, 43, 58, 86, 89, 99
Punchdrunk, 7, 13, 25, 29, 30, 55, 56, 57, 58, 60, 61, 64, 67, 78, 79, 85, 86, **89–99**, 105, 106, 131, 203
  Faust, 90, 93, 94
  The Masque of the Red Death, 57, 94, 95, 203
  Sleep No More, 13, 57, 68, 93, 96
  see also Barrett, Felix; Doyle, Maxine

Ramachandran, V.S., 6, 15, 16, **18–19**, 52, 199, 200
  and Carrie-Armel, K., 199, 200
  and Hubbard, E.N., 6, 15, 16, **18–19**
Rodaway, Paul, 199, 200
Rosenberg, David, 25, 86, **100–11**
Royal Court Theatre, 61, 71
  4.48 Psychosis, 61, 86, 156, 159
  Far Away, 71, 86, 152
Russian Formalists, the, 8, 34, **37–9**, 56
  see also Bakhtin, Mikhail; Brik, Osip; Shklovsky, Victor
Ruthrof, Horst, 198, 202
  see also intersemiotic analysis

Sealey, Jenny, 61, 87, **160–71**
Scarry, Elaine, 8, 21, 23, 25, 26, 47, 201
Schneider, Rebecca, 27, 200
Shklovsky, Victor, **37–8**, 57, 69, 75, 80
  and experiential form, 37–8, 57

Ostranenie, 37
  see also defamiliarization
  'special perception', 69, 78
  see also Russian Formalists, the
Shunt, 7, 25, 29, 30, 55, 56, 58, 60, 61, 85, **100–11**, 131
  Amato Saltone, 106, 110
  The Ballad of Bobby Francois, 58, 104
  Dance Bear Dance, 54, 104, 105, 107
  Shunt Lounge, 106
  The Tennis Show, 105
  Tropicana, 57, 103, 108, 109
Sierz, Aleks, 31
site-specific, 56, 57, 92, 203–4
  'site-sympathetic', 56, 61, 85, 92, 203–4
Synaesthesia, 13–21
  defined, 4, 13–14
  diagnostic features, 15–16, 18–19
  drug-induced, 19
  medical condition, 6, 7
  neurocognitive, 6, 9
  in scientific research, 6, 15–20
  and synkinaesia, 19
(Syn)aesthetics defined, 4, 13–33
  and the body, 21–4, 62–9
  chthonic, 5, 6, 22, 27, 34, 36, 41, 42, 44, 50, 52, 57, 67, 68, 72, 74, 79, 85, 86, 198, 200, 202
  hybridity, 4, 5, 8, 26, 29–30, **55–62**
  imagination, 5, 17, 18, 21–2
  and the ineffable, 16, 20, 21, 32, 36, 55, 70, 72, 73, 98, 119, 142, 193
  and the intangible, 20, 21, 24, 36, 49, 67, 68, 70, 72, 98, 133, 151, 163, 188
  making sense/*sense*, 14, 20, 21, 22–4, 36, 39, 42, 45, 51, 55, 61, 62, 66, 68, 69, 70, 74, 78, 80, 148, 200
  noetic, 16, 18, 21, 34, 36, 48, 57, 68, 69, 70, 72, 74, 78, 85, 86
  in performance practice, **54–81**

(Syn)aesthetics defined – *continued*
  *play*text, 4, 5, 18, 24, 32, 37–44,
    46–50, **69–80**, 86, 198, 132–59
  and the political, 24, 28, 61, 64, 67,
    70, 86, 109, 112, 121, 125, 126,
    132–43, 194, 202
  somatic/semantic, 4, 14, 17–19,
    20, 36, 39, 42, 45, 47, 60, 61,
    75, 78, 80, 87
  (syn)aesthetic-sense, **20**, 21, 25,
    36, 39, 45, 49, 55, 61, 67, 68,
    71, 72, 73, 74
  visceral-verbal, 4, 18, 24, 31–2, 37,
    40, 42, 49, **69–80**, 85, 137, 200

Taussig, Michael, 15, 52, 198, 201
Theatre de Complicite, 3, 29, 30
  Street of Crocodiles, 3
Tozer, Kathy, 147

Van Campen, Cretien, 15, 19,
  198, 200
visceral defined, 1, 197

Wagner, Richard, 4, 203
Wallace, Naomi, 7, 22, 24, 31, 32,
  42, 55, 64, 67, 69, 73, 74, 79,
  86, **132–43**, 202
  Slaughter City, 134
  Things Of Dry Hours, 70, 74, 86,
    132–43
  Trestle at Pope Lick Creek, 75
Werner, Heinz, 19
Wilson, August, 134
Wilson, Robert, 30
writerly, 4, 22, 31, 39, 40, 46,
  47, 48, 49, 53, 69, 70, 73,
  79, 80
  defined, 197